BENCHMARKING
GLOBAL
MANUFACTURING
UNDERSTANDING INTERNATIONAL
SUPPLIERS, CUSTOMERS,
AND COMPETITORS

BENCHMARKING GLOBAL MANUFACTURING
UNDERSTANDING INTERNATIONAL SUPPLIERS, CUSTOMERS, AND COMPETITORS

Jeffrey G. Miller

Arnoud De Meyer

Jinichiro Nakane

With contributions by other members of the Global Manufacturing Futures Project, including:

Kasra Ferdows, INSEAD, France

Kee Young Kim and Dae Ryun Chang, Yonsei University, Korea

Norma J. Harrison, University of Technology, Australia

Ta Huu Phuong and Lee Soo Ann, National University of Singapore, Singapore

Seiji Kurosu, Waseda University, Japan

Jay Kim, Boston University, United States

Lawrence M. Corbett, Victoria University, New Zealand

Rafael Arana, Miguel Leon, and Carlos Cosio, IPADE, Mexico

Haruki Matsuura, Kanagawa University, Japan

BUSINESS ONE IRWIN
Homewood, Illinois 60430

© RICHARD D. IRWIN, INC., 1992

Sponsoring editor: Jean Marie Geracie
Project editor: Rebecca Dodson
Production manager: Irene H. Sotiroff
Jacket designer: Michael Finkelman
Designer: Maureen McCutcheon
Art manager: Kim Meriwether
Compositor: BookMasters, Inc.
Typeface: 11/14 Palatino
Printer: Book Press, Inc.

Library of Congress Cataloging-in-Publication Data

Miller, Jeffrey G.
    Benchmarking global manufacturing / Jeffrey G. Miller, Arnoud De
Meyer, Jinichiro Nakane; with contributions by other members of the
Global Manufacturing Futures Project, including Kasra Ferdows . . .
[et al.].
        p.   cm. — (The Business One Irwin/APICS series in production
management)
    Includes index.
    ISBN 1-55623-674-3
    1. Manufactures. I. De Meyer, Arnoud. II. Nakane, Jinichiro,
1934– . III. Title. IV. Series.
HD9720.5.M55 1992
338.4′767—dc20   91-45517
                                                              CIP

Printed in the United States of America
1 2 3 4 5 6 7 8 9 0 BP 9 8 7 6 5 4 3 2

PREFACE

This book is written for you if you are a manufacturing executive challenged with creating a "factory of the future" that reflects the best practices of world-class manufacturers and the strategic needs of your own organization. You are an important translator of your company's strategic intentions into operational action. As you define the factory of the future for your business, you must make sure that manufacturing reinforces the mission of other functions, and that other functions provide support for manufacturing. Concurrently, you must help the whole organization, from president to blue-collar worker, to prepare itself to compete in an environment in which employees, customers, suppliers, and competitors come from widely different national, cultural, and political backgrounds.

The book will help you by providing benchmarking information about global manufacturing companies along with the tools to help you use the information in your organization. The Benchmarking information we refer to includes measures of the performance, strategies, operations, and ways of thinking of large, geographically diverse, global manufacturing companies. It can provide a perspective on the nature and impact of the cultural gaps that must be crossed to build the global manufacturing corporation.

The benchmarking information and tools described in the book will enable you to directly compare your own manufacturing strategies, operations, and plans to manufacturers from various industry groups (including your own) in the major industrial nations. These tools will also help you to gain insights about how to improve your operations, achieve consensus across functions and at different organizational levels, and revitalize traditional benchmarking efforts.

This book is the result of a unique, 10-year study, the Global Manufacturing Futures Project. At Boston University in 1981, the Project began to systematically gather data about the performance, status, and plans of global manufacturers in a variety of industries. Since then, we have learned much about global manufacturing that is new and exciting—particularly in the area of benchmarking, and the special branch we call *strategic benchmarking*. Strategic benchmarking involves learning how to improve the manufacturing operations of one company by systematically comparing perceptions about it to those of similar companies from around the world. These perceptions are used to challenge assumptions, to make large groups of people think rigorously and strategically about what they are doing, and to provide a basis for consensus about needed future improvements.

The book is not intended to be a compendium of all the information that you might require to understand manufacturing in different parts of the world. Various international guides provide information on geography, travel, currency, language, culture, and business customs. Government agencies from the various countries usually provide data on employment, rates of pay, imports and exports, capacity utilization, investment policies, and so on. A number of other private and governmental sources are available that project economic growth, national income, and specific market data. Where appropriate, we will reference these sources.

Nor have we addressed manufacturing everywhere. In focusing on the United States, Western Europe, and Japan, and providing a glimpse into such diverse countries as Mexico, Korea, Australia, New Zealand, and Singapore, we hope to provide information about some of the largest and most interesting players on the industrial scene. The most noticeable omissions from industrialized nations are manufacturers from Eastern Europe and the former Soviet Union. When we started gathering our data, these countries were not a factor in global manufacturing trade, certainly not in terms of the capitalist model. If these countries can maintain and build on their new-found freedom to compete, future editions of this book will certainly include information about them. We also expect to expand our coverage of South America in the future.

The book is divided into three sections. The first section familiarizes the reader with the changing international environment in manufacturing and the strategic benchmarking methods we have developed. Examples of the ways leading companies have employed strategic benchmarking techniques are provided, along with background on the manufacturing strategy model used as a basis for it.

The second part of the book contains five chapters that outline, compare, and interpret the results of surveys of over 1,000 manufacturing business units from around the world. These companies have all been administered the Strategic Benchmarking Questionnaire as a part of the Manufacturing Futures Project. Chapters 5, 6, and 7 interpret the results and trends in the United States, Western Europe, and Japan, respectively. Chapter 8 provides an overview of the results from five countries that encircle the Pacific Rim and also offers some very different views of manufacturing. Chapter 9, Factories of the Future, compares and contrasts the findings.

The chapters in Part 2 are written by authors from each of the countries or regions described. These chapters reflect manufacturing thought in these countries as interpreted by residents deeply involved in shaping it. This second section of the book can be used as a general introduction to current manufacturing thought in various parts of the world and thus provides a broad basis for comparison for the reader attempting to benchmark his or her own company against them.

The third part of the book contains data tables that readers can use to benchmark their operations against the responses of the American, Japanese, and European companies included in the international study. These data are broken down into five industry groups and by country. In the case of Western Europe, these data are presented by region (e.g., the Latin countries), as well as by industry. In addition to their use for detailed benchmarking, these data have been in high demand by market researchers, consultants, and others seeking information on potential and developing manufacturing customers and markets. We have promised the companies that participated in the study with us that we would not reveal their identities. Pseudonyms are used when particular companies are described. The data for individual companies participating in the Manufacturing Futures Survey are not disclosed in Part 3. Means of groups are used in these tables to protect the identity of survey participants.

The final appendix to Part 3, Appendix D, contains a glossary of important terms we use in the book. These terms are set boldface in the text. The language of modern manufacturing is continually evolving, as are the meanings of the words and acronyms used. The glossary we have developed is an attempt to capture the best meanings that were translatable into a variety of tongues and cultures in 1990, when the data were collected.

This book has depended on far more than the usual number of people to produce. First mention must go to the founding members and the sponsors of the Manufacturing Futures Project research teams from various countries and at various points in time. The project was started at Boston University in 1981 by Jeff Miller with the sponsorship of Arthur Andersen & Co. In 1983, the scope of the study was expanded to include Europe, in collaboration with Professors Tom Vollmann and Kasra Ferdows at INSEAD, and with Professor Jinichiro Nakane at Waseda University in Japan. In 1983, the Manufacturing Roundtable at Boston University assumed sponsorship of the project in collaboration with the Program on Management of Technology & Innovation at INSEAD and the Waseda Systems Science Institute. The survey portion of the project was administered in these countries on an annual basis through 1988, and again in 1990. We plan to administer the survey again in 1992 and 1994.

Over time, various individuals have made important contributions to the survey strategy and methodology. Miller developed the original survey model, much of which remains intact today. In 1983, Vollmann and Ferdows assumed responsibility for a major rewrite of the questions that enabled them to be translated more readily into eight languages, and expanded the scope of the survey. In 1986, Aleda Roth at Boston University and Arnoud De Meyer at INSEAD brought needed rigor to the statistical methods used to analyze and interpret the data and made some important changes in survey methodology. In 1988, Kee Young Kim of Korea, Louis Ta of Singapore, and Norma Harrison of Australia contributed important thinking to the overall strategy for the survey. Jinichiro Nakane's ideas from his Manufacturing 2001 Project were a critical stimulus to this rethinking. More recently, Jay Kim at Boston University, Seiji Kurosu at Waseda, Lawrie Corbett from New Zealand, and Miguel Leon and Carlos Cosio and their team from Mexico have developed important new improvements. Each "generation" has built on the foundation provided by others, left its mark, and grown in its understanding of international management and what it means to work in a true alliance. Perhaps most importantly, we have had fun, and hope to have more.

Our colleagues and especially the deans of the institutions involved deserve our thanks for their patient support and understanding. It takes special circumstances to enable a large, far-flung group of scholars to continue their work over such a sustained period. By encouraging our work when international studies were less popular, helping to organize conferences, covering classes while we traveled, and providing invaluable commentary and financial support, our colleagues and leaders created these circumstances.

Among the many other contributors who have worked on the project over the last 10 years, several deserve special mention for their role in developing this book. Linda Angell organized and merged the many individual reports developed in Part 2 of the book. Thean-jeen Lee at Boston University and

Kenneth Bonheure at INSEAD developed the tables in Part 3. Janet Bond Wood applied her considerable editorial skill to integrating chapters written by different authors with different native tongues into one coherent book, and offered invaluable advice and commentary to the authors. Professor Jay Kim coauthored Chapters 5 and 9, helped with Chapter 3, and used his charm and quick wit to help other authors see where they could make improvements in Part 2. Mr. Gerry Angeli of Eastman Kodak and Mr. Alan Biddle of Eli Lilly reviewed the book and offered many useful comments. Mr. Angeli was especially helpful in providing new insight into the art of product teardowns.

Finally, but foremost, we wish to thank the thousands of manufacturing executives and companies who participated in the Manufacturing Futures Survey over the years. It is to them that we dedicate this book.

Jeffrey G. Miller, Boston, United States
Arnoud De Meyer, Fontainebleau, France
Jinichiro Nakane, Tokyo, Japan

CONTENTS

About APICS

APICS, the educational society for resource management, offers the resources professionals need to succeed in the manufacturing community. With more than 35 years of experience, 70,000 members, and 260 local chapters, APICS is recognized worldwide for setting the standards for professional education. The society offers a full range of courses, conferences, educational programs, certification processes, and materials developed under the direction of industry experts.

APICS offers everything members need to enhance their careers and increase their professional value. Benefits include:

- Two internationally recognized educational certification processes—Certified in Production and Inventory Management (CPIM) and Certified in Integrated Resource Management (CIRM), which provide immediate recognition in the field and enhance members' work-related knowledge and skills. The CPIM process focuses on depth of knowledge in the core areas of production and inventory management, while the CIRM process supplies a breadth of knowledge in 13 functional areas of the business enterprise.
- The APICS Educational Materials Catalog—a handy collection of courses, proceedings, reprints, training materials, videos, software, and books written by industry experts...many of which are available to members at substantial discounts.
- *APICS The Performance Advantage*—a monthly magazine that focuses on improving competitiveness, quality, and productivity.
- Specific industry groups (SIGs)—suborganizations that develop educational programs, offer accompanying materials, and provide valuable networking opportunities.
- A multitude of educational workshops, employment referral, insurance, a retirement plan, and more.

To join APICS, or for complete information on the many benefits and services of APICS membership, **call 1-800-444-2742** or **703-237-8344**. Use extension 297.

PART 1

BENCHMARKING

CHAPTER 1

THE NEW NEED TO KNOW

At a recent international conference in Tokyo a young Japanese executive asked, "Why do we need to hear about manufacturing in the United States or Europe? They no longer have anything to teach us." The response of a more seasoned Japanese executive was that around the world, there is a new and increasingly critical need to know about manufacturing wherever it is done. This view is held by a large number of firms from the United States, Europe, and Japan that benchmark global manufacturing, that is, that compare themselves to the best manufacturing firms around the globe.

The need for a manufacturing firm to know about how the best manufacturing firms do what they do has been present for many years. In Japan, for example, the need to know has traditionally been driven by the desire to develop the technical and managerial know-how to rebuild its industry in the postwar era. Numerous benchmarking trips to manufacturers in the United States and Europe helped Japanese manufacturers acquire this knowledge.

Today, however, the need to know has acquired a new urgency and a different thrust. The new need to know in Japan is driven by fresh global market opportunities and competitors, and by the need to learn about new technologies and managerial techniques from the large number of countries that now have the capability to produce them. The need to revisit America and Europe is also important as leading manufacturers in these countries regain and sharpen their competitive edge.

The new need to know in Western Europe is driven by political change, as well as by global market opportunity. Traditional structures and organizations have been more than upset by the drive toward European integration and the ongoing turmoil in what was the Soviet Union and Eastern Bloc. For example, the common wisdom on logistics or factory location has become obsolete. European manufacturers are looking at other manufacturers to see how organizations, manufacturing methods, plant locations, and logistical arrangements can be adapted to their changing situations.

In the United States, the new need to know arises from a change in perspective. Manufacturers once content to focus on the single largest homogeneous market in the world have found that their strategies must change. Unlike the postwar era, other countries can now successfully compete for the U.S. market. But at the same time, more manufacturers have found significant offshore markets they can exploit, as evidenced by the growth in export sales from 6 percent of U.S. GNP in 1968 to 14 percent in 1988. American manufacturing companies are now looking widely to see how others have succeeded in developing these international opportunities.

Global manufacturing, once confined to a few large consumer goods and oil and chemical multinationals, has spread as manufacturing companies in one industry after another begin to operate on an international scale. The ongoing wave of mergers and acquisitions has contributed to this trend. Acquisitions such as Philips' European white goods appliances division by Whirlpool, Firestone by Japan's Bridgestone, RCA's TV division by France's Thompson, Japan's Banyu by Merck, or Britain's ICL by Fujitsu have created new companies that operate manufacturing and distribution networks around the world.

The modern manager from any country must be equipped to deal with industrial customers, suppliers, and competitors from different parts of the world. In Japan today, for example, 17 percent of the sales of large Japanese manufacturing companies in the Manufacturing Futures Survey are to offshore customers, 6.1 percent of sales are produced in a foreign location, and 10 percent of purchases are from offshore suppliers. In 1989, the large companies in the U.S. Manufacturing Survey exported 21 percent of their sales, purchased 16 percent of their materials and components from offshore suppliers, and produced 11 percent of their output in offshore locations. In Europe, the formation of the European Community (EC) has resulted in the rationalization of century-old industrial structures. Patterns of demand and competition that were stable for 40 years have suddenly become obsolete, resulting in more foreign sales and manufacturing within and outside Europe. Figure 1–1 shows that the trend toward a more global base for manufacturing companies continues.

In order for modern manufacturing managers to expand their perspective to match the global scale of operations, they must understand and value foreign ways of thinking and planning. They must be able to respect the differences in opinion

FIGURE 1–1
Percent Change in Offshore Activities, 1989–1992

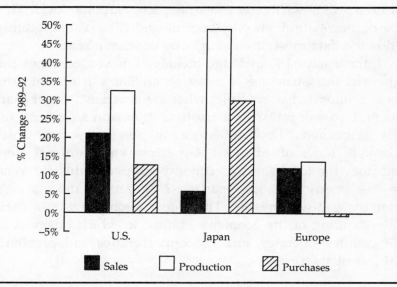

of counterparts around the world in order to work toward common goals. But the few firms that have educated their employees to understand foreign ways and practices have found that even this is not enough. They must go beyond understanding to the discovery of new ways of management that capitalize on the synergies obtainable by combining the best from different cultures. In the words of one senior manager from an American firm, "Here we are planning manufacturing in Japan and most of our people have never even learned how to get a passport before, let alone how to get the most from the combination of our approach and Japanese culture and technology. We've got to learn more about operating abroad fast."

WHAT DO YOU NEED TO KNOW?

The modern manager must understand global manufacturing at three levels: the infrastructural level, the cultural level, and the strategic/operational level. The need for structural and cultural knowledge, while important, is hardly new. This book emphasizes knowledge at the strategic/operational level. This type of information has only become widely available recently and is the data most often sought by benchmarkers.

Infrastructural knowledge includes knowledge about the industrial infrastructure of a nation and how it fits into the world economy. For example, what are a nation's major markets and growth rates? Primary industries and suppliers? Industrial locations? Levels of education? Average wages? Costs of capital? Rates of inflation? Regulatory environment? How well does the local service economy support industry? Who are the primary trading partners? What are the nation's strengths and weaknesses? This information provides a basic understanding of the economic context in which business in that country operates, and the opportunities and problems that industry faces.

A number of sources provide infrastructural information. Good English language overviews are provided by publications such as the *Economist* series on international business[1] or Michael Porter's *Competitive Advantage of Nations.*[2] More in-depth knowledge, as well as ongoing information about changes in the manufacturing infrastructure of a nation, can be obtained from the United Nations and other economic sources listed in the references at the end of this chapter.

Cultural information includes the history, language, politics, beliefs and values, norms of behavior, social problems, and the customs of the people. It provides the foundation for understanding the motives and likely reactions of individuals or groups from that society, as well as information about how to behave in various business and social transactions. For most nations, a variety of sources exist for information about culture and society. Some particularly good references for the United States, Japan, and Western Europe are also included in the references section at the end of this chapter.

Strategic/operational knowledge includes information about the ways in which manufacturers from various nations attempt to compete. Of course, every individual organization differs in its strategy and operations to some extent. The only way to obtain specific knowledge about another manufacturing company is to study it carefully as an entity. The study of the variation in strategy and operations among the firms in a country prepares one to relate to them as individual customers, suppliers, and competitors.

In benchmarking the strategy and operations of manufacturing firms, three issues deserve special attention: strategic intent, the results of continuous experimentation, and continuous improvement. The strategic intent of an organization identifies the ends it seeks to achieve, experimentation helps to identify the appropriate means by which these ends can be

---

[1]"European Trends," The *Economist*, Intelligence Unit, London, 1991.
[2]Michael Porter, *Competitive Advantage of Nations* (New York: Basic Books, 1990).

served, and continuous improvement identifies how an organization continually refines its approaches to meet its ends.

Strategic Intent

A critical issue for any firm is where to emphasize improvement. Overall improvement is desired, but scarce resources, customer needs, and market forces dictate that choices be made with respect to where investments in time, money, and management attention will be made.

The essence of an effective manufacturing strategy is the development of those competitive capabilities that will best position the firm for sustainable competitive advantage. The top competitive capabilities, or priorities, guide the choices that complete the formulation of an effective manufacturing strategy and continuous improvement plan. To the extent that these priorities are future oriented, they signify the strategic intent of the firm. Table 1–1 suggests how the strategic intent of nations differs within the manufacturing community; for example, the Americans put more emphasis on price, the Japanese emphasize reliability and the ability to change designs and introduce new products quickly, and the Europeans emphasize performance and delivery. Each region values different aspects of quality, such as conformance, reliability, or performance. Data about strategic intent are even more useful when gathered at the industry level. For example, an

TABLE 1–1
Top Five Competitive Priorities in the Next Five Years

Europe	Japan	U.S.
Conformance quality	Product reliability	Conformance quality
On-time delivery	On-time delivery	On-time delivery
Product reliability	Fast design change	Product reliability
Performance quality	Conformance quality	Performance quality
Delivery speed	Product customization	Price

electronics firm can judge the extent to which its strategic intent is different from its competitors in a particular part of the world. Part 3 of this book contains the data for such comparisons.

Continuous Experiments

As the competitive pace has quickened over the last 10 years, manufacturing firms and their suppliers around the world have become marvelously creative in achieving their strategic intent. A large number of the innovations in the management process have risen and fallen in popularity quickly, proving that they were mere fads. Others have had staying power—as the early adopters of these innovations have learned by virtue of persistence how to make them useful in manufacturing.

The manufacturing firm wishing to sustain a competitive advantage is in a precarious position when choosing which innovations to employ. If the firm waits too long to see if an innovation is successful, it may lose valuable time compared to those who adopt it earlier. On the other hand, early adoption of a number of different innovations can lead the firm into chaos—unable to distinguish between good and bad and caught in a spiral of confusion.

We view the current manufacturing world as conducting a large number of market tests, or experiments, on these management tools and process innovations. The data in this book identify which innovations have been most and least useful to those who have tried to use them (see Table 1–2, for example). Though this sort of benchmarking information must be used cautiously, it does provide insights that can stimulate deeper thought about the nature of the latest management developments.

Further, the data in Part 3 identify the innovations and action plans in which manufacturing firms plan to invest through 1992. These investments indicate the changes that are being made to manufacturing strategies as firms around

TABLE 1–2
Which Innovations Pay Off in the United States?

Highest *Pay-off*	1. Interfunctional work teams 2. Manufacturing reorganization 3. Statistical quality control (SQC) 4. Linking manufacturing to business strategy 5. Just-in-time (JIT) ⋮
Lowest *Pay-off*	22. Flexible manufacturing systems 23. Integrating information systems across functions 24. Integrating information systems in manufacturing 25. Activity-based costing (ABC) 26. Robotics

the world build their own versions of the "factory of the future."

Continuous Improvement

The reader will soon be convinced by the information in Part 2, if he or she is not already, that manufacturers around the world are locked in a battle of continuous improvement. Rates of learning and the speed of implementation of new ideas will be the key issues in manufacturing in the coming decade. Manufacturing is about hard work and implementation. Companies that understand how to shorten the time to implement new ideas successfully, and who can learn from that implementation how to speed up future improvement efforts, will sow the seeds for the next generation of competitive success.

Figure 1–2 shows the average rates of improvement on a number of variables critical to manufacturing firms from the United States, Japan, and Europe. By benchmarking your company against these average rates of improvement, you can begin to gauge whether your firm is learning at world-class rates. This book contains information and tools that will assist you to do this comparison for your company.

FIGURE 1–2
Performance Improvements, 1988–1989

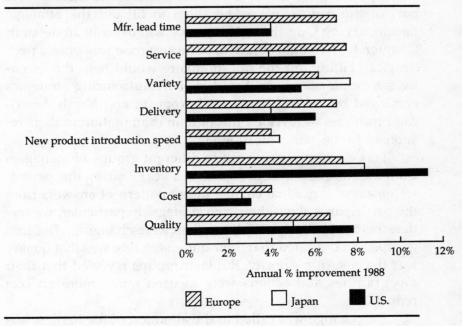

STRATEGIC BENCHMARKING

As we worked with the participants in the Manufacturing Futures Project over the years, we began to see that they needed more than data and information such as Figure 1–2 and Table 1–2 to help them become good international partners and competitive manufacturers. Benchmarking is not enough. An organization needs to be prepared to effectively use benchmarking information. In other words, along with a greater "new need to know" about global manufacturing comes an equally necessary "new need to know" about internal beliefs and perceptions. We discovered what the Chinese General/Philosopher Sun Tzu knew over 2,000 years ago: "Know your enemy *and yourself*, and in a hundred battles you will never be in peril."

Our experience with a large Southeast Asian producer of consumer goods illustrates the wisdom of this Chinese philosopher. The company had organized a series of executive development programs for its senior middle management. As part of the course, we asked them to fill out the Strategic Benchmarking Questionnaire that we will describe in detail in Chapter 3. The initial objective of this exercise was purely pedagogical: Filling out the questionnaire would help the executives to understand the breadth of manufacturing strategies employed by leading Japanese, European, and North American producers as revealed through an examination of their responses to the same questionnaire.

This exercise was offered to different groups of managers from the company over the space of a year. During this period, we observed a gradual change in the pattern of answers from the participants about their own strategy. In particular, we saw that their priorities and action plans were changing. The first groups had indicated that their understanding was that quality was the clear top priority. But later groups revealed that their top priorities and actions were focused much more on cost reduction.

This change was called to the attention of the highest levels of management, who became quite concerned. They knew that intervening events had included some actions that might have led the rest of the organization to conclude that there had been a shift in strategy, but this had not been their intention. They became concerned that they could not build the core competencies that were necessary to ensure long-term competitiveness if their employees' perceptions of the manufacturing priorities could change so considerably over such a relatively short period of time. Was top management encouraging short-term thinking at the expense of long-term competitiveness? As a result of its concern, the company initiated a task force to redefine its manufacturing strategic intent and its portfolio of key actions and programs, as well as to consider ways to project more long-term thinking into the organization.

In addition to short-term thinking, we have observed two other major impediments that stand in the way of converting external benchmarking data into effective internal action in companies around the world. The first is that cross-functional business processes such as product development, TQC (total quality control), and materials management are becoming more central to competing in global markets than activities that take place within the confines of a specific functional area. The emphasis on cross-functional business processes, in conjunction with the trend toward more participative management styles, means that there is a need for more consensus in the organization about what to benchmark and what to do with the results.

Evidence of the new importance of cross-functional activities is provided by Table 1–2, which shows that cross-functional teams have become the single most important management tool in American firms. The story of a Silicon Valley electronics manufacturer provides a vivid illustration of why. Figure 1–3 shows the different levels of importance attributed to product development speed by various functions in this company. Data were collected as part of a workshop on manufacturing benchmarking. After these functionally based perceptions on the importance of various capabilities were fed back to the workshop, the organization was able to identify a

FIGURE 1–3
Importance of Time to Market across Functions

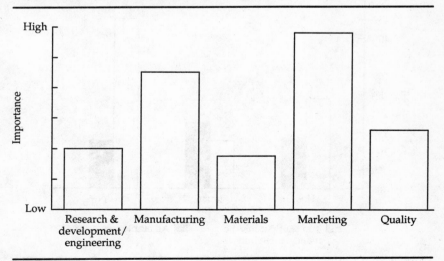

key problem blocking their progress. The research and development organization and the materials organizations were not convinced that shorter times to market were worthwhile goals, even though the company's customers had explicitly told marketing that this was a critical variable for them. After recognizing the internal problem, the company leadership reassigned responsibilities so that the functions could work collectively to gather information on world-class product development and implement the results.

The popularity of participative management approaches around the world is indicated by Figure 1–4, which shows that a significant percentage of manufacturers broadly share information about strategies, goals, and plans with all employees. In the United States, the percentage increased from 3 percent of companies in 1984 to nearly 14 percent in 1990. This trend toward participation requires that more of an organization's employees understand the direction of the company in order to use external benchmarking information.

FIGURE 1–4
Respondents (percent) Who Share Knowledge of Goals and Plans, 1990

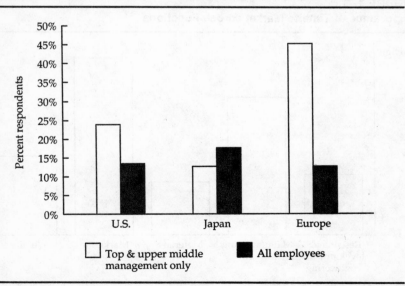

For example, a large American auto manufacturer, Autoco, was one of several dozen firms where we piloted the use of our Manufacturing Survey instruments as tools for strategic benchmarking. (This company will be discussed at greater length in Chapter 4.) The company had already benchmarked a number of its processes and as a result had initiated numerous programs such as just-in-time (JIT), statistical quality control (SQC), total quality control (TQC), and so on. Unfortunately, none of them seemed to be having an impact. Sixty employees of the company, from blue-collar to senior executive, representing most of the functional areas including manufacturing, engineering, marketing, and finance, filled out our Strategic Benchmarking Survey (described in detail in Chapters 3 and 4) in order to gauge their understanding of the key manufacturing capabilities, operations, and policies of the company, the importance and cross-connections between the new programs, and their standing vis-à-vis other leading manufacturing firms.

An analysis of the results indicated that the relative importance ascribed to quality improvement by each level, from blue-collar worker to line management to staff, varied significantly. Though the company had been working hard to share information, knowledge, and responsibility, it is clear from the data that the company's information campaign and its smorgasbord of improvement efforts had produced more confusion than consensus. The company did not have a clearly shared strategic intent or knowledge base and thus, the various levels of employees reacted to external benchmarking information in very different ways. Blue-collar workers thought top management overreacted to external data on quality, while top management thought workers overreacted to data on costs.

The lack of an integrated manufacturing strategy prevents organizations from using functional benchmarking effectively. Without such a strategy, it will not be clear what the most important processes are to be benchmarked. It is as important to understand *what* the right processes are to benchmark and *why* they need to be done exceptionally well, as it is to observe *how* the particular process is done in a world-class environment.

A final case in point is from the shipbuilding industry. A large firm followed the practice of many today by initiating a continuous improvement program. As a result, individuals were sent to various other firms to observe and benchmark their business practices. One individual came back with a glowing report of the process a competitor used to plan and control materials throughout the entire engineering, preproduction, and construction process. The result? Nothing happened. Why? There was no agreement within the corporation about basic strategic priorities, and little understanding of how the information gathered could be used to make ship construction a strategic weapon.

A year later, after an intensive period of self-examination and widespread discussion and consensus building on the strategy, a team of top managers returned to a competitor's shipyard to observe the materials process. This time, the results were accepted and the implementation process was initiated with the widespread support of the managers in the enterprise. "The difference," said the director of planning, "was that the second time we all knew what the right questions were to ask and why they were important to us."

We hope that this book helps to make a similar difference for those who use it by providing an approach, along with relevant data, for strategic benchmarking. By strategic benchmarking, we mean both the internal process of gaining consensus on what is important to ask and why, and the external process of determining the processes and yardsticks that are most useful.

HOW TO USE THIS BOOK

There are three ways in which this book can be profitably used, depending on the depth of the reader's need to know. The first way is to use it purely as a resource book on modern manufacturing as it is practiced in the major industrialized countries and a sampling of smaller manufacturing nations.

Readers at this level will concentrate their efforts on Part 2 of the book, which contains descriptions of manufacturing issues, plans, operations, and current trends in these countries. If such readers are interested in market research and want to gain a closer view of an industry within each region, or more detail on one region, then they can refer to the detailed data in Part 3.

Readers at the second level will be more interested in benchmarking their firms against the sample of global manufacturers described in Parts 2 and 3. These readers will want to read all three sections of the book in detail and fill out the Manufacturing Futures Strategic Benchmarking Questionnaire. These readers will learn from the examples in Chapter 4 how to compare the questionnaire responses to those of a reference group, and will then go on to make these comparisons using Chapters 5–9 and Part 3.

Readers who will gain the most from this book will use it to start the process of strategic benchmarking in their plants, divisions, groups, or companies. Like the second level of readers, they will read Chapters 2 and 3 and respond to the questionnaire. But then they will go on to use all or parts of the benchmarking questionnaire in Chapter 3 and the examples in Chapter 4 with a large group of people with whom consensus is important. Rather than compare responses to the Strategic Benchmarking Questionnaire and Part 3 alone, they will use the internal responses along with the external data (preferably with the aid of a trained facilitator) to drive a process that leads toward internal consensus on the key questions and their motivation, and toward continuous benchmarking and improvement.

REFERENCES

Abegglen, J. C., and G. Stalk. *Kaisha*. New York: Basic Books, 1985.
De Meyer, A.; J. Miller; J. Nakane; and K. Ferdows. "Flexibility, the Next Competitive Battle." *Strategic Management Journal*, Winter 1989.

"European Trends," The *Economist* Intelligence Unit. London. 1991.

Evans, P.; Y. Doz; and A. Laurent. *Human Resource Management in International Firms.* New York: Macmillan, 1989.

Ferdows, K., ed. *Managing International Manufacturing.* New York: Elsevier North Holland, 1989.

Ferdows, K.; J. Miller; J. Nakane; and T. Vollmann. "Evolving Manufacturing Strategies: Projections into the 1990s." *International Journal of Operations and Production Management* (special issue), January 1987.

Hofstede, G. *Culture's Consequences.* Beverly Hills, Calif.: Sage Publications, 1980.

————. "The Interaction between National and Organizational Value Systems." *Journal of Management Studies* 22, no. 4, 1985, pp. 347–57.

Imai, M. *Kaizen.* New York: Random House, 1986.

Ishikawa, K. *What Is TQC?* New York: Prentice Hall, 1985.

Laurent, A. "The Cross-cultural Puzzle of International Human Resource Management." *Human Resource Management* 25, no. 1, 1986, pp. 91–102.

Nakane, J., and R. Hall. *Flexibility Manufacturing Battlefield in the 90s.* Chicago: Association for Manufacturing Excellence, 1990.

OECD Economic Outlook, Organization for Economic Cooperation and Development, Publications Service. Paris. 1991.

OECD Economics Surveys, Organization for Economic Cooperation and Development, Publications Service. Paris.

Porter, M. *Competitive Advantage of Nations.* New York: Basic Books, 1990.

Ward, P., and T. Vollmann. "Mapping Manufacturers Concerns and Action Plans." *International Journal of Operations and Production Management* 8, no. 6, 1988.

CHAPTER 2

STRATEGIC BENCHMARKING

Strategic benchmarking provides strategic data and information that can be compared to similar information from other global manufacturing companies. If done well, it also organizes a company's priorities and plans and builds consensus on the areas to benchmark in more detail.

Strategic benchmarking is just one of several ways to benchmark—an activity that varies depending on whether a product, a process, customer needs or global strategies are being compared. This chapter describes different types of benchmarking and the ways they relate to strategic benchmarking, and outlines the procedural steps you can follow to initiate your own benchmarking of global manufacturing.

There are four kinds of benchmarking:

1. Product benchmarking.
2. Functional or process benchmarking.
3. Best practices benchmarking.
4. Strategic benchmarking.

Of these, the oldest and most widely used is product benchmarking. Functional or process benchmarking has been more recently elevated into a high science by the Xerox Corporation, and even more recently, best practices benchmarking has emerged. Strategic benchmarking provides a way to start viewing manufacturing from a global perspective, and to ensure that the customer needs are brought into the benchmarking process.

Product Benchmarking

Product benchmarking is the long-standing practice of carefully examining or "tearing down" another manufacturer's product. The other manufacturer whose product is being benchmarked might be a direct competitor, a sister division, or a manufacturer of noncompetitive products that employs similar technologies or who attempts to satisfy the same customers.

The range of information that is available from product benchmarking is quite broad. Sometimes, competitive products are "reverse engineered" so that they can be copied, or so that competitor costs can be estimated. More often, products are benchmarked so they can be compared to new product concepts or prototypes. In this way, a company can evaluate the current and future strengths and weaknesses of alternative designs. But in sophisticated companies, the reasons for product benchmarking are much more subtle. They see product benchmarking as an opportunity to understand new or different ways to respond to customer needs, design products, or manufacture.

Mr. Gerry Angeli of Eastman Kodak puts it this way: "Most people think of product benchmarking as elementary biology—just dissecting a frog to see what the parts are. But product benchmarking is really like archeology, where you dig to find out as much as you can about a civilization." For example, by observing how fastening mechanisms are designed, and by reassembling a product after tearing it down, an engineer can tell a great deal about how design for assembly is being implemented. By using the product as directed, a designer may be able to deduce how a competitor thinks about design tradeoffs, or learn a new way to satisfy a customer's need from a noncompetitive product.

Functional Benchmarking

Functional (or process) benchmarking became popular in the mid 1980s. It is similar to product benchmarking in that the aim was to learn how to improve. Indeed, functional benchmarking has remained closely associated with the continuous improvement and competitiveness movements that began in the United States in the early 1980s. Functional benchmarking differs from product benchmarking in two important ways. First, the focus of comparison is on a functional process such as order entry, assembly, testing, product development, setting up a machine, or shipping. Second, functional benchmarking cannot be done effectively without the permission of the company being benchmarked.

In order for an organization to do any successful functional benchmarking, it must admit that it is possible that someone, either a competitor or a noncompetitor who performs a similar process, is doing it better. Thus, pride, coupled with the not-invented-here syndrome, can be a barrier. The problems associated with changing a culture so that it looks externally for comparison rather than to its own history of improvement are significant. Xerox has become one of the most widely referenced companies because of its success in developing a strong continuous improvement culture built around functional process benchmarking.

Best Practices Benchmarking

"There's a crucial difference between best practices and the benchmarking lots of companies do. Benchmarkers usually study nonpareils in particular functions—that is, what can our shipping department learn from L. L. Bean. GE looks less for nuts and bolts and more for attitudes and management behavior."[1]

Best practices benchmarking goes beyond functional benchmarking by focusing the search on management practices, as well as the elements of the functional processes that managers oversee. While the basic approach to benchmarking as practiced by GE has much in common with functional benchmarking, its emphasis on management practices remains a subtle but important difference. For example, a best practices

---

[1]"How Jack Welsh Keeps the Ideas Coming at GE," *Fortune,* August 13, 1991.

benchmarking study might focus on how successful companies compensate their employees at different levels. Compensation practices can be expected to influence all functional processes and the attitudes of the employees involved in them. Best practices benchmarking attempts to take benchmarkers beyond comparisons of work procedures to the comparisons of management practice that enable good processes to perform in an exceptional manner.

Strategic Benchmarking

Strategic benchmarking begins by comparing the strategic intent of the company doing the benchmarking with that of the company being benchmarked. In other words, it begins with the proposition that one must consider how the competitor defines success before attempting to know the secrets of that success. Conversely, the benchmarking company must also develop a sound notion of its own strategic intent before it knows what to look for in others.

Unlike functional or best practices benchmarking, strategic benchmarking is based on the assumption that clear communication, a common language, consensus, and participation are required before the fruits of other types of benchmarking can be enjoyed. By insisting that the benchmarking company define its own strategic intent, the strategic benchmarking procedure attempts to ensure that there is consensus within the organization about basic directions and key performance measures, as well as a common language for describing them. This, in turn, helps to ensure that the company focuses on the most critical functions to benchmark, and knows why it is doing the benchmarking and how to use the information it receives.

For example, if the company doing strategic benchmarking is customer oriented, then strategic benchmarking will help assure that the best practices or processes being benchmarked are the ones that contribute the most to the satisfaction of customer needs. The strategic intent of a customer oriented manufacturer will embody key customer values. Companies whose strategy focuses on developing or exploiting certain core competencies will find that strategic benchmarking orients employees toward practices and processes that contribute to this end.

All benchmarking techniques are justified by the possibility of improvement that they offer. But strategic benchmarking looks first at the *ends* to be accomplished. In contrast, functional and best practices benchmarking compare the *means* first (the processes, functions, or practices), and then tries to sort out whether they are better in achieving some end.

Functional benchmarking conducted in isolation can trigger actions that do not contribute to the company's strategic goals. Continuous improvement assumes that the addition of a large number of small improvements will lead to large improvements in the performance of the firm, as shown in Figure 2–1. But this is only true if all continuous improvement activities are guided by the same intent and properly coordinated. A coordinated strategic direction can be used to evaluate the contribution of improvements learned from benchmarking, and help avoid the possibility that one improvement activity cancels the effect of another, as suggested by Figure 2–2.

FIGURE 2–1
What We Hope Continuous Improvement Provides

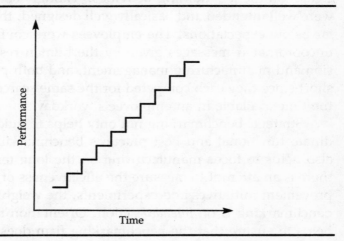

FIGURE 2–2
What Happens When Continuous Improvements Cancel Each Other

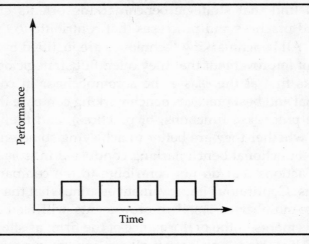

To illustrate how good ideas can cancel each other, consider the following. Both the human resources division and the manufacturing management of a large firm carried out benchmarking activities independently. Each implemented a new program to improve the skill level of employees. The human resource management division designed an ambitious training program, while manufacturing management decided to increase on-the-job training activities. Though both programs were well intended and basically well designed, the result was far below expectations. The employees were confused by the uncoordinated messages given by the human resources division and manufacturing management, and both programs fell short since they each competed for the same scarce resource—the time available in an employee's workday.

Strategic benchmarking not only helps to guide and coordinate functional and best practices benchmarking efforts, it also helps to focus manufacturing on the long term. Though there is an attempt to measure the effectiveness of existing improvement initiatives and experiments, the weight in strategic benchmarking is on *intended* action. Orientation to the future helps to ensure that the benchmarking firm does not end up simply following the efforts of the leaders, and maximizes the probability that it will benchmark those activities that are critical in developing competencies needed in the future.

Though the progression from product benchmarking through strategic benchmarking has been evolutionary in practice, that does not mean that one form is superior to another. Each has a somewhat different focus and provides different types of information. Product benchmarking usually produces the most hard and specialized information, but it is also the information that has the shortest life. In many fast-moving industries, product benchmarking is viewed as an attempt to document history. At the other end of the spectrum, we see that strategic benchmarking provides more general, soft, and somewhat less precise information. Future-oriented and dynamic information is prone to be more uncertain than history. However, our studies have shown that this information has a relatively long life in terms of its applicability and use. Figure 2–3 illustrates the tradeoffs between the four methods.

We recommend that all four forms of benchmarking be employed; all have something valuable to offer. But we also recommend that strategic benchmarking be the first of the methods to be used, since it provides a context and a rationale that enhances the effectiveness of the other three types of benchmarking. Without strategic benchmarking as a first step, other benchmarking efforts are blind.

Strategic benchmarking is the job of senior manufacturing management, top management, and management from other functions. In contrast, best practices benchmarking is usually done by staff, functional benchmarking by specialists involved in the business processes, and product benchmarking by product specialists. In setting the course of strategic benchmarking, top management provides critical leadership for the firm.

BENCHMARKING PROCEDURES

If you are convinced that benchmarking in its various forms has value, then you must move from thought to action. How can I do strategic benchmarking? How do I follow up with

FIGURE 2–3
Precision versus Life of Benchmarking Information

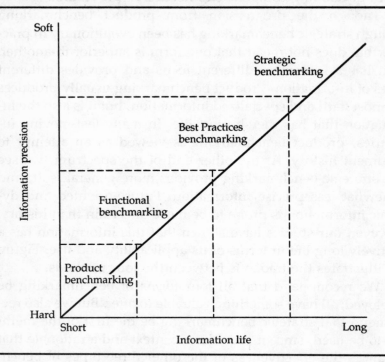

other types of benchmarking once the first step has been finished?

Over the 10 years since the Manufacturing Futures Project was begun, we have developed the following procedure for strategic benchmarking. This procedure has been used by over 100 companies, divisions, and plants around the world as a way to initiate or give new energy to their global manufacturing benchmarking and strategy development efforts, while achieving high levels of employee participation and continuous improvement. Though some adaptation of this procedure is necessary in almost every instance, the basic steps remain fairly constant.

1. *Fill out the Strategic Benchmarking Questionnaire.* The strategic benchmarking instrument from the Manufac-

turing Futures Survey has provided a useful common framework for individual companies to use. Its advantages are that it has been tested in a number of companies, and a company's response to it can be compared to those of thousands of other global manufacturers over the last ten years. The questionnaire can be found at the end of Chapter 3. When used to challenge conventional thinking and generate consensus, the questionnaire (or portions of it) is administered privately to individuals (often a relevant sample) within a business unit.

2. *Compare the questionnaire results internally.* The strategy of the company must be developed internally before it can be benchmarked externally. The mean responses of the individuals in an organization to the questionnaire provide but a rough cut of the range of initial perceptions that exist about strategic intent, future actions, and experiences. In a sense, the responses to the questionnaire are probes into what people think is the truth on matters such as where the company is going, what its priorities are, what has worked operationally, and what should be done in the future.

3. *Benchmark the results against those of other reference groups.* The responses to the questionnaire are then compared to the data in Part 3 of this book so that the benchmarking company can compare the perceptions of its employees to those of other reference groups by industry and/or location. For example, the numeric ratings of the competitive capability portion of the questionnaire that show the firm's strategic intent and its strength on various competitive capabilities can be compared to the strength and strategic intent of various country and industry groups from Part 3.

4. *Analyze the differences.* If the questionnaire is answered by a large number of people from an organization, it is usually helpful to break this large group into small groups of 6 to 10 individuals to analyze the

responses. These groups can assess the similarities and differences between their own responses and the responses of the relevant comparison groups. It is easy at this stage to rationalize that things really aren't all that different by focusing on similarities. While similarities are often interesting, a focus on differences generally provides a more stimulating view. Why are we going in a different direction from competitors in another part of the world? Why aren't we as successful in improving or in using new technology? These questions challenge the organization to look past its own nose in considering the future.

The objective of this comparison is not to convince an organization to change its strategy and actions so that it is similar to that of others, though that occasionally happens in companies that do not have well-developed manufacturing and business strategies. Rather, the purpose is to clarify and challenge the internal view, to validate it, and to increase everyone's understanding of its implications. It is also critical in helping to identify the important functional and best practices benchmarking targets.

The consensus of the organization emerges only after the group and its leadership have been able to clarify and validate strategic intent and experience, and to vigorously agree to pursue a particular manufacturing strategy. This is accomplished by carefully analyzing major differences to the responses to the questionnaire internally, and then attempting to resolve them. In small groups, this can be done by hand scoring the responses with hash marks on a master copy of the questionnaire. With larger groups, computer analysis is helpful. In either instance, it is useful to have a trained facilitator to assist the groups in debating the meaning of the varied responses and coming to some consensus. In those instances where consensus is not possible (or where the variance in responses is high), more information from outside the organization or group is typically needed. This is the juncture at which the questions to be asked in external product, functional, or best practices benchmarking are defined.

5. *Find the questions.* The comparison and assessment in earlier steps should help in the construction of a common language and generate more questions for functional, product, or best practices benchmarking. For example, one company concluded after this strategic benchmarking exercise that its human resources strategy needed to be examined in further benchmarking efforts. It became clear from the analysis that in order for the company to achieve its aims, it either had to retrain existing personnel or find new people. No amount of functional benchmarking could overcome the deficit in human capital that existed. Another company concluded that it did not have an adequate grasp of customer needs after the first four steps. It developed a plan to find out by analyzing the activities of some of its customers.

   The analysis should also help to clarify the units of analysis that are most appropriate for further discussion and investigation. Not infrequently, this process shows that one plant or portion of an organization should be focused on particular customers. This will be shown when several groups respond to the questionnaire in different ways that can only be explained by different customer demands. The strategic issue this raises is whether these different groups should be legitimized, and if so, how the key benchmarking questions for these subgroups differ from one another.

6. *Search out the answers.* As noted at the outset, the practice of strategic benchmarking is just the first step. Follow-on steps are dictated by the results of the analysis and may include further functional or best practices benchmarking, the generation of a more complete manufacturing and business strategy, or reorganization

to better focus business activity and further continuous improvement. Functional and best practices benchmarking can be accomplished by carrying out the steps outlined in the next list.

7. *Act on the results and start over again.* At this point in the process, the external benchmarking efforts of the organization will have produced a "to do" list that represents the key findings from benchmarking at all four levels that have been identified. The organization will also find that it has a much better awareness of other organizations, and that it will be much better prepared as it repeats this set of steps.

Steps 5 and 6 in the Strategic Benchmarking Procedure suggest that you determine the key questions to be asked about global manufacturing, and then search out the answers through best practices, functional (process), and product benchmarking procedures. The following procedure, developed by *Industry Week* in their survey of current practices, indicates the typical steps you can expect to follow in answering the questions raised by strategic benchmarking as you progress to best practices or functional benchmarking:[2]

1. *Identify the function (or best practice) to benchmark.* This, of course, is what strategic benchmarking helps you to do.
2. *Identify the best-in-class company.* Once a key competitive factor is identified, find out who performs best on that factor or who does the process best. Seek their cooperation in a benchmarking study.
3. *Identify the key performance variables to measure and collect the data.* Agree with the cooperating company on the type of data being collected to ensure that accurate comparisons are made.

---

[2] "Benchmarking," *Industry Week*, November 5, 1990.

4. *Analyze and compare the data to what happens in your own company.* In addition to collecting quantitative data, describe key management approaches that differ between the companies, and identify the critical success factors.

5. *Project future performance levels of the benchmarked company.* Since it will take some time to adapt many of the techniques used by the benchmarked company, project what that company's performance level will be in five years, then plan to match or exceed that level in your own business.

6. *Establish functional goals.* The team of benchmarking specialists presents its final recommendations to top management on ways in which the organization must change to reach the new goals.

7. *Communicate benchmark findings.* Senior management shares portions of the benchmarking study's findings with employees to build support and enthusiasm for new strategies.

8. *Develop action plans.* The benchmarking team develops specific action plans for each objective.

9. *Implement specific actions and monitor progress.* After a period of time, collect data to determine new performance levels. Adjustments in the action plans should be made if the goals are not being met, and problem-solving teams formed to work out snags.

10. *Recalibrate benchmarks.* Over time, the benchmarks should be reevaluated and updated to ensure that they are based on the latest data and best targets.

The organization, circumstances, and context of the company doing the strategic and other types of benchmarking dictate the particular adaptations to the procedures described above, and the types of outcomes that may result. In Chapter 4, we illustrate with actual examples how strategic benchmarking has been combined with other benchmarking procedures in different circumstances to help an organization understand how to compete in a global manufacturing environment.

REFERENCES

"Benchmarking." *Industry Week*, November 5, 1990.

Camp, R. C. *Benchmarking: The Search for the Best Industry Practices that Lead to Superior Performance.* Milwaukee: ASQC Quality Press, 1989.

"How Jack Welsh Keeps the Ideas Coming at GE." *Fortune*, August 13, 1991.

Skinner, W. "The Focused Factory." *Harvard Business Review*, September/October 1974.

Walleck, A. S.; J. D. O'Halloran; and C. A. Leader. "Benchmarking World Class Performance." *The McKinsey Quarterly*, January 1991.

CHAPTER 3

A TOOLKIT FOR STRATEGIC BENCHMARKING

In the previous chapter, we outlined an approach to strategic benchmarking and described how it could be used in conjunction with more traditional methods for functional and best practices benchmarking. In the next chapter, we will illustrate how major international manufacturing companies have done strategic benchmarking to improve their global manufacturing posture. The purpose of this chapter is to introduce you to the toolkit we have provided for doing strategic benchmarking. This toolkit includes the questions in the Strategic Benchmarking Questionnaire, located at the end of this chapter, and the data in Part 3 of this book.

THE STRATEGIC BENCHMARKING QUESTIONNAIRE

The Strategic Benchmarking Questionnaire contains questions that a manufacturing business unit must answer about itself and its competitors to begin the benchmarking process. The questions fall into five categories, as shown in Figure 3–1.

Strategic Business Unit Identification

The definition of the unit of analysis to be assessed is extremely important in benchmarking. Setting the boundaries for a business is in itself a critical strategic decision in an organization.

FIGURE 3–1
Categories of Strategic Benchmarking Questions

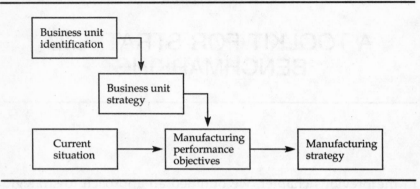

Once the boundaries are determined, the scope of the markets, technologies, customer base, and product lines to be considered becomes apparent.

At one level, it is easy to determine the relevant business unit for benchmarking. The simple approach is to identify existing organizational units as business units. But organizational boundaries don't always match up very well with strategic boundaries. The strategic boundaries of a business are defined by the value chain of supply, design, manufacture, sales, and distribution activities that are oriented to serving a particular customer base. Organizational boundaries, in contrast, are determined by internal trade-offs, rather than by market forces. For example, several companies in the Manufacturing Futures Survey have organized themselves with one manufacturing division whose responsibility is to produce products for a wide range of product or sales divisions. For them, identifying the business unit as a division would be a mistake. The strategic unit of analysis for any one customer set would cut across several divisions.

The questions in the Strategic Benchmarking Questionnaire that are focused on identifying the strategic business unit enable an individual, or a group of individuals, to sort out the particular unit of analysis for which they are answering. In the case where the survey is administered to many people, these questions also help to ensure that everyone is dealing with the same unit of analysis.

Current Situation

The second category of questions in the Strategic Benchmarking Questionnaire asks about the current status of a number of factors in the strategic business unit. The answers to these questions form a baseline against which the practices and performance of other companies can be compared, as well as a baseline for comparison with future performance and plans. Included in this group of questions are those that delve into current business practices such as the extent of information sharing, the perceived competitiveness on various factors compared to competition, and the degree to which previous investments in new technology or practice have paid off.

Business Unit Strategy

The category of questions oriented toward discovering the strategy of the business unit is composed of three parts. The first part seeks to ascertain the growth goals of the business unit. The second develops an understanding of the current profile of global manufacturing, procurement, and selling activities. The third focuses on identifying the importance of various capabilities that the business unit intends to use in meeting customer needs and distinguishing itself from its competitors. Of particular importance here is the identification of the core capabilities that form the basis of the strategic intent of the business unit.

Table 3–1 lists the capabilities examined in the Strategic Benchmarking Questionnaire.

Manufacturing Performance Objectives

The fourth category of questions in the questionnaire pertains to the performance objectives or targets of the manufacturing function. The list in Table 3–2 illustrates the range

TABLE 3–1
Key Competitive Capabilities

	Price
Low price	Ability to profit in price competitive markets
	Flexibility
Design change	Ability to make rapid changes in design
New products	Ability to introduce new products quickly
Volume change	Ability to make rapid volume changes
Mix change	Ability to make rapid product mix changes
Broad line	Ability to offer a broad product line
	Quality
Conformance	Ability to offer consistently low defect rates
Performance	Ability to provide high-performance products or product amenities
Reliable products	Ability to provide reliable/durable products
	Delivery
Fast delivery	Ability to provide fast deliveries
On-time delivery	Ability to make dependable delivery promises
	Service
After-sales services	Ability to provide effective after-sales service
Support	Ability to provide product support effectively
Distribution	Ability to make product easily available
Customize	Ability to customize products and services to customer needs

of possible performance measures considered. The importance attributed to each of these targets should be influenced by the business unit strategy as well as other, shorter term considerations. A key issue for you to consider is whether the performance targets are consistent with the strategic directions of the organization.

Other aspects of manufacturing performance that are elicited by these questions identify planned changes in the global

TABLE 3–2
Manufacturing Performance Objectives

Improve conformance quality (reduce defects)	Improve ability to make rapid volume changes
Reduce unit costs	Reduce break-even points
Improve safety record	Raise employee morale
Reduce manufacturing lead time	Increase environmental safety and protection
Increase capacity	Reduce capacity
Reduce procurement lead time	Increase product or materials standardization
Reduce new product development cycle	Improve labor relations
Reduce materials costs	Improve white-collar productivity
Reduce overhead costs	Increase range of products produced by existing facilities
Improve direct labor productivity	Meet financial shipping goals
Increase throughput	Improve pre-sales service and technical support
Reduce number of vendors	Improve after-sales service
Improve vendor quality	Change culture of manufacturing organization
Reduce inventories	Improve interfunctional communication
Increase delivery reliability	Improve communication with external partners
Increase delivery speed	Reduce set-up/changeover times
Improve ability to make rapid product mix changes	

distribution of manufacturing, sourcing, and selling activities, and future cost structure goals.

Manufacturing Strategy

A manufacturing strategy identifies the actions that the manufacturing organization intends to take in the future in order to support the business unit strategy and meet manufacturing

TABLE 3–3
List of Manufacturing Action Plans

Giving workers a broad range of tasks and/or more responsibility (job enlargement/enrichment)	Integrating information systems in manufacturing
Activity-based costing	Integrating information systems across functions
Manufacturing reorganization	Reconditioning physical plants
Worker training	Just-in-time (JIT)
Management training	Robots
Supervisor training	Flexible manufacturing systems
Computer-aided manufacturing (CAM)	Design for manufacture
Computer-aided design (CAD)	Statistical quality control (SQC)
Value analysis/product redesign	Closing and/or relocating plants
Interfunctional work teams	Quality circles
Quality function deployment	Investing in improved production-inventory control systems
Developing new processes for new products	Hiring in new skills from outside
Developing new processes for old products	Linking manufacturing strategy to business strategy

performance objectives. Table 3–3 describes the various action programs and investment alternatives that are included as indicators of change in manufacturing policy in the Strategic Benchmarking Questionnaire. The list is in no way exhaustive. It is the result of the careful analysis of what is potentially common to many companies and industries. Over the years, this list has been the most dynamic of all the items in the Strategic Benchmarking Questionnaire.

The answers to this set of questions are important for two reasons. Internally, it is important to determine whether the manufacturing strategy is consistent with manufacturing performance objectives and business unit strategy. Externally, it is useful to benchmark your own manufacturing strategy against companies with different strategies and situations around the world.

Figure 3–2 shows the five categories of questions in the questionnaire and indicates by question number which ques-

FIGURE 3–2
Strategic Benchmarking Questionnaire Questions by Category

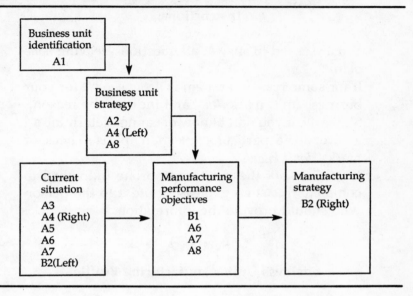

tions are included in which category. You will note that the answers to several questions appear in several categories. This is because we have combined various subquestions in order to make it easier to actually use the questionnaire.

THE MANUFACTURING FUTURES DATABASE

To use the data you provide in the Strategic Benchmarking Questionnaire, you need comparative data against which to benchmark your responses. Part 3 of this book has the responses that nearly 500 strategic business units from the U.S., Japan, and Europe supplied from January to June of 1990. These data are essential to the strategic benchmarking toolkit.

In the next chapter, we will describe several ways in which manufacturing companies have used the toolkit elements to strategically benchmark their operations against those of global manufacturers.

STRATEGIC BENCHMARKING QUESTIONNAIRE*

Instructions:

- You are asked to answer all questions, leaving *none* blank.
- If for some reason, an item is *not applicable* for your business unit, mark *"na"* and indicate the reason.
- Similarly, if you don't know or cannot determine an answer for a particular question, please check or mark *"don't know."*
- If you are not the most appropriate individual to complete this survey, please route it to the person who should provide the information.

Section A

Business Unit/Manufacturing Profile

1. Check the one category that *best describes the business unit* for which you are responding to this questionnaire.

    Company [ ]   Division [ ]   Plant [ ]
    Other (specify) _____ [ ]

    a) What is the name of the business unit for which you are responding?

    _____

    b) What is the name of the highest level parent corporation of which the business unit is a part?

    _____

    c) Please briefly describe in your own words the primary product/family of the business unit:

    _____

    _____

---

*Copies of the Strategic Benchmarking Questionnaire can be obtained with permission to use from the Boston University Manufacturing Roundtable.

2. Check the one item below that best characterizes the basic thrust of the business unit over the next two years. (Check one only):

   Build market share   [ ]
   Hold (defend) market share   [ ]
   Harvest (maximize cash flow, sacrifice share)   [ ]
   Withdraw (prepare to exit the business)   [ ]

3. Check the one item below to describe who understands the goals, strategies, and overall business plans in the business unit. (Check one only):

   Top management only   [ ]
   Top and most middle management   [ ]
   Top and some middle management   [ ]
   Every manager and supervisor   [ ]
   Every manager, supervisor, and worker   [ ]

4. For each competitive ability on the next page, circle the number *on the left-hand scale* that indicates the degree of importance to the business unit in competing in the marketplace over the next *5 years.*

   On the right-hand scale, circle the number that best describes your business unit's current competitive strength relative to the *competitor who does it best in your industry.*

   Note that if you circle the same number for two items, it implies that the two items are of approximately equal importance.

Degree of Importance	COMPETITIVE ABILITIES	Degree of Strength Relative to Best Competitor
Very Unimportant *Very Important*		*Weak* *Strong*
	PRICE Ability to profit in **price competitive** markets	
1 2 3 4 5 6 7		1 2 3 4 5 6 7
	FLEXIBILITY Ability to make **changes in design** and introduce **new products** quickly	
1 2 3 4 5 6 7		1 2 3 4 5 6 7
1 2 3 4 5 6 7	Ability to make **rapid volume changes**	1 2 3 4 5 6 7
1 2 3 4 5 6 7	Ability to make **rapid product mix changes**	1 2 3 4 5 6 7
1 2 3 4 5 6 7	Ability to **offer a broad product line**	1 2 3 4 5 6 7
	QUALITY Ability to offer **consistent quality**	
1 2 3 4 5 6 7		1 2 3 4 5 6 7
1 2 3 4 5 6 7	Ability to provide **high-performance** products or **product amenities**	1 2 3 4 5 6 7
1 2 3 4 5 6 7	Ability to provide **reliable** products	1 2 3 4 5 6 7
	DELIVERY Ability to provide **fast deliveries**	
1 2 3 4 5 6 7		1 2 3 4 5 6 7
1 2 3 4 5 6 7	Ability to make **dependable delivery promises**	1 2 3 4 5 6 7
	SERVICE Ability to provide effective **after-sales service**	
1 2 3 4 5 6 7		1 2 3 4 5 6 7
1 2 3 4 5 6 7	Ability to **provide product support** effectively	1 2 3 4 5 6 7
1 2 3 4 5 6 7	Ability to make product **easily available** (broad distribution)	1 2 3 4 5 6 7
1 2 3 4 5 6 7	Ability to **customize** products and services to customer needs	1 2 3 4 5 6 7

5. Please provide the following information about the business unit for the *last complete fiscal year*. (Please make sure the figures below correspond to the business unit—company, division, or plant):

Date of end of last fiscal year
for which data is reported . . . . .   _____

Annual sales revenues (millions
of dollars) . . . . $ _____ millions

Pre-tax return on assets (Pre-tax
profit ÷ Total assets) × 100 . . . .   _____ %

Net pre-tax profit (Pre-tax
profit ÷ Total sales) × 100 . . . . .   _____ %

Research and development
expenses as percent of sales
revenue (R&D expense ÷ Total
sales) × 100% . . . .   _____ %

Growth rate in unit sales
(percent increase over previous
fiscal year) . . . .   _____ %

Market share of primary
product . . . .   _____ %

Capacity utilization (last fiscal
year) . . . .   _____ %

Number of plants in the
business unit . . . .   _____

Number of employees
(manufacturing and
non-manufacturing) . . . .   _____

Manufacturing direct labor
employees . . . .   _____

Manufacturing indirect labor
employees . . . .   _____

SIC (Standard Industrial
Classification) code for business
unit . . . .   _____

6. Estimate the past and future manufacturing costs of the business unit as a percent of total sales in each period.

	1989 Actual Mfr. Cost/Sales	1992 Actual Mfr. Cost/Sales	199X Planned Mfr. Cost/Sales
Manufacturing costs as a percent of sales	_____ %	_____ %	_____ %

7. Estimate the past and future manufacturing cost structure for your business unit. (Percentages should add to 100%.)

	1989 Actual Cost Structure	1992 Actual Cost Structure	199X Planned Cost Structure
Materials	_____ %	_____ %	_____ %
Direct labor	_____ %	_____ %	_____ %
Energy	_____ %	_____ %	_____ %
Manufacturing overhead	_____ %	_____ %	_____ %
TOTAL	**100%**	**100%**	**100%**

8. Estimate for the business unit the recent (1992) and projected (199X) percentages of U.S. (domestic) and foreign sales, production, and purchases.

	Sales		Production		Purchases	
	1992	199X	1992	199X	1992	199X
U.S. (domestic)	____	____	____	____	____	____
Foreign	____	____	____	____	____	____
	100%	100%	100%	100%	100%	100%

9. In the following list, we are asking you to mentally construct an index for each manufacturing performance indicator. Note that the beginning of 1989 is the baseline year so that the value of the index in 1989 = 100. You are to ascribe a relative value to each indicator listed below as of the beginning of the year 1992. An index number greater than 100 indicates improvement, while an index of less than 100 indicates a decline in performance. For example, if your unit costs declined by 25% from the beginning of 1989 to the beginning of 1992 (an improvement), then the index for 1992 would be 125. Similarly, if your on-time delivery worsened by 30% from 1989–1992, then the 1992 index would be 70.

Indicator	1992 Relative Index (1989 = 100)
Overall quality as perceived by customers (higher = better)	_____
Average unit production costs for typical product (lower = better)	_____
Inventory turnover (higher = better)	_____
Speed of new product development and/or design change (faster = better)	_____
On-time delivery (more reliable = better)	_____
Equipment changeover (or set-up) time (faster = better)	_____
Market share (higher = better)	_____
Profitability (higher = better)	_____
Customer service (higher = better)	_____
Manufacturing lead time (smaller = better)	_____
Procurement lead time (smaller = better)	_____
Delivery lead time (faster = better)	_____
Variety of products producible by manufacturing (more = better)	_____

Section B

Manufacturing's Objectives and Plans

1. Indicate the relative importance in the next two years of each of the following objectives for the business unit manufacturing function by circling the appropriate number on the scale. Note that if you circle the same number for two items, that implies the two items are equally important.

	Not Important	Critically Important
Improve conformance quality (reduce defects)	1 2 3 4 5 6 7	
Reduce unit costs	1 2 3 4 5 6 7	
Improve safety record	1 2 3 4 5 6 7	
Reduce manufacturing lead time	1 2 3 4 5 6 7	
Increase capacity	1 2 3 4 5 6 7	
Reduce procurement lead time	1 2 3 4 5 6 7	
Reduce new product development cycle	1 2 3 4 5 6 7	
Reduce materials costs	1 2 3 4 5 6 7	
Reduce overhead costs	1 2 3 4 5 6 7	
Improve direct labor productivity	1 2 3 4 5 6 7	
Increase throughput	1 2 3 4 5 6 7	
Reduce number of vendors	1 2 3 4 5 6 7	
Improve vendor quality	1 2 3 4 5 6 7	
Reduce inventories	1 2 3 4 5 6 7	
Increase delivery reliability	1 2 3 4 5 6 7	
Increase delivery speed	1 2 3 4 5 6 7	
Improve ability to make rapid product mix changes	1 2 3 4 5 6 7	
Increase ability to make rapid volume changes	1 2 3 4 5 6 7	
Reduce break-even points	1 2 3 4 5 6 7	
Raise employee morale	1 2 3 4 5 6 7	
Maximize cash flow	1 2 3 4 5 6 7	
Increase environmental safety and protection	1 2 3 4 5 6 7	
Reduce capacity	1 2 3 4 5 6 7	
Increase product or materials standardization	1 2 3 4 5 6 7	

	Not Important	Critically Important
Improve labor relations	1 2 3 4 5 6 7	
Improve white-collar productivity	1 2 3 4 5 6 7	
Increase range of products produced by existing facilities	1 2 3 4 5 6 7	
Meet financial shipping goals	1 2 3 4 5 6 7	
Improve pre-sales service and technical support	1 2 3 4 5 6 7	
Improve after-sales service	1 2 3 4 5 6 7	
Change culture of manufacturing organization	1 2 3 4 5 6 7	
Improve interfunctional communication	1 2 3 4 5 6 7	
Improve communication with external partners	1 2 3 4 5 6 7	
Reduce set-up/changeover times	1 2 3 4 5 6 7	
Other (describe) _____	1 2 3 4 5 6 7	

2. Please follow these instructions for the items listed on the next two pages:
   (a) *On the far-left side,* check those activities, tools, or programs that were given a significant degree of emphasis in the past two years.
   (b) *On the second left-hand scale,* indicate the extent of pay-off resulting from these activities, tools, or programs in the past two years, but only if they have been emphasized in the last two years.
   (c) *On the right-hand scale,* indicate the relative degree of emphasis the business unit will place on each type of activity, tool, or program over the next two years.

Check If Emphasized within Last Two Years	Relative Pay-Off Last Two Years		PROGRAMS/ ACTIVITIES	Degree of Emphasis Next Two Years	
	Little Pay-Off	*Great Pay-Off*		*Little Emphasis*	*Great Emphasis*
[ ]	1 2 3 4 5 6 7		Giving workers a broad range of tasks and/ or more responsibility	1 2 3 4 5 6 7	
[ ]	1 2 3 4 5 6 7		Activity-based costing	1 2 3 4 5 6 7	
[ ]	1 2 3 4 5 6 7		Manufacturing reorganization	1 2 3 4 5 6 7	
[ ]	1 2 3 4 5 6 7		Worker training	1 2 3 4 5 6 7	
[ ]	1 2 3 4 5 6 7		Management training	1 2 3 4 5 6 7	
[ ]	1 2 3 4 5 6 7		Supervisor training	1 2 3 4 5 6 7	
[ ]	1 2 3 4 5 6 7		Computer-aided manufacturing	1 2 3 4 5 6 7	
[ ]	1 2 3 4 5 6 7		Computer-aided design	1 2 3 4 5 6 7	
[ ]	1 2 3 4 5 6 7		Value analysis/ product redesign	1 2 3 4 5 6 7	
[ ]	1 2 3 4 5 6 7		Interfunctional work teams	1 2 3 4 5 6 7	
[ ]	1 2 3 4 5 6 7		Quality function deployment	1 2 3 4 5 6 7	
[ ]	1 2 3 4 5 6 7		Developing new processes for new products	1 2 3 4 5 6 7	
[ ]	1 2 3 4 5 6 7		Developing new processes for old products	1 2 3 4 5 6 7	
[ ]	1 2 3 4 5 6 7		Integrating information systems in manufacturing	1 2 3 4 5 6 7	
[ ]	1 2 3 4 5 6 7		Integrating information systems across functions	1 2 3 4 5 6 7	
[ ]	1 2 3 4 5 6 7		Reconditioning physical plants	1 2 3 4 5 6 7	
[ ]	1 2 3 4 5 6 7		Just-in-time	1 2 3 4 5 6 7	
[ ]	1 2 3 4 5 6 7		Robots	1 2 3 4 5 6 7	

Check If Emphasized within Last Two Years	Relative Pay-Off Last Two Years		PROGRAMS/ ACTIVITIES	Degree of Emphasis Next Two Years	
	Little Pay-Off	*Great Pay-Off*		*Little Emphasis*	*Great Emphasis*
[ ]	1 2 3 4 5 6 7		Flexible manufacturing systems	1 2 3 4 5 6 7	
[ ]	1 2 3 4 5 6 7		Design for manufacture	1 2 3 4 5 6 7	
[ ]	1 2 3 4 5 6 7		Statistical quality control	1 2 3 4 5 6 7	
[ ]	1 2 3 4 5 6 7		Closing and/or relocating plants	1 2 3 4 5 6 7	
[ ]	1 2 3 4 5 6 7		Quality circles	1 2 3 4 5 6 7	
[ ]	1 2 3 4 5 6 7		Investing in improved production-inventory control systems	1 2 3 4 5 6 7	
[ ]	1 2 3 4 5 6 7		Hiring in new skills from outside	1 2 3 4 5 6 7	
[ ]	1 2 3 4 5 6 7		Linking manufacturing strategy outside to business strategy	1 2 3 4 5 6 7	

CHAPTER 4

HOW GLOBAL MANUFACTURERS USE STRATEGIC BENCHMARKING

The following examples of how global manufacturers have used strategic benchmarking illustrate the wide range of industrial and cultural situations in which it can be employed. They also provide insight into how the procedures described in the previous chapters can be adapted to large- versus small-group settings, to settings where there is more than one business unit (plant, division), and how to use the technique with suppliers and customers.

The first example, Chemco, describes how a small group of senior manufacturing executives from one of Europe's top chemical companies used strategic benchmarking and the information from Part 2 of this book to set corporate direction. The Autoco example provides more detail on the example cited in Chapter 1 of how an organization used strategic benchmarking techniques to achieve strategic alignment between various levels in the organization. Because a large group was involved in this example, it also illustrates the simple statistical techniques that can be used to best analyze the data in Part 3 of the book. These first two examples follow the steps outlined in Chapter 2 fairly closely, and are organized in this fashion.

The third example of strategic benchmarking, Diversico, shows how a diversified corporation used the technique to initiate manufacturing strategic planning and divisional and corporate strategic planning, and illustrates the flexibility of the technique in engaging customers and suppliers in the strategic benchmarking process.

EXAMPLE 1: CHEMCO

Chemco is a large European chemical company producing pharmaceuticals, bulk chemicals, agrichemicals, and specialty chemicals. Its specialty chemical division was confronted with an increasing variety of products, in ever smaller volumes. The production equipment available to this division was mainly designed for relatively stable bulk products, and thus, maladapted to frequent changes in mix, volume, and due date.

The senior manufacturing manager had set as an objective to run the equipment as a flexible manufacturing system, even though large amounts of idle and downtime would result until the equipment could be truly adapted to flexible use. Increasing the flexibility of operations was put high on the formal list of strategic priorities for the business. Manufacturing management, supervisors, and the skilled labor force had been trained in the management of flexibility. But the senior manufacturing manager suspected that the commitment to flexibility had not been completely internalized by his employees and colleagues in other functions, and that much of the commitment to flexibility did not go beyond slogans and exhortation.

The senior manager of operations decided to use the results of the European Manufacturing Futures Survey (Chapter 6) to test the commitment of his team to flexible operations. He invited the researchers in the Manufacturing Futures Project to come and lead a discussion with a group of 12 managers from manufacturing, logistics, and process engineering. From that point, the process employed in Chemco paralleled the steps in strategic benchmarking outlined in Chapter 2.

1. Fill Out the Strategic Benchmarking Questionnaire

2. Compare the Results Internally

In preparation for the meeting, the 12 managers filled out the parts in the benchmarking questionnaire related to competitive priorities, performance, and past and future action plans. A first analysis of their answers revealed that these 12 people, who were in daily contact with each other and were supposed to have a common view on the manufacturing strategy, had in fact very different views. Overall, the average responses showed that the 12 managers agreed, at least on paper, with the priority that flexibility had to take over other competitive abilities. But agreement did not exist about the evaluation of the company's position vis-à-vis important competitors, and even less agreement existed about the action plans that had to be pursued to achieve flexibility.

It was clear that the differences in opinions were largely functional, divided between the process engineering group, the logistics people, and the manufacturing operations managers. The first group had a clear commitment to investment in flexible technologies and considered the factory to be leading among its competitors. The second group thought delivery dependability was more important than flexibility, and wanted to have investments and actions that facilitated the attainment of this goal. The third group adhered to the flexibility goal, but emphasized investments in cost-reduction programs.

The differences angered the participants at the presentation and generated a heated debate. It became clear that the disagreement was fueled by the conflicting performance measures (and maybe the egos) of each subgroup. Though the managers in charge of operations could support the drive for flexibility, they were evaluated on capacity utilization and cost efficiency. The logistics people were constantly pressured by customer purchasing departments to shorten the delivery delays, so to them, a dependable response to delivery requests was much more important than a flexible response.

3. Benchmark the Results Against Those of Other Reference Groups

4. Analyze the Differences

The reference group had been defined before the presentation. The criteria were that companies included in the group needed to utilize activities and process technologies close to those of Chemco. (The reference group was comprised of the European basic industries and consumer industries subgroups identified in Part 3, Appendix C.) The senior manufacturing manager of Chemco had insisted that the group should not be limited to companies that could be considered competitors. He also wanted it to include companies with technologies resembling Chemco's but that delivered to other markets. He was convinced that some industries had been confronted with similar market challenges at earlier dates and had already developed some practices that might be useful to Chemco.

The comparison of Chemco's responses to the benchmarking questionnaire with those of the reference groups suggested that Chemco was lagging in manufacturing information systems and control procedures. Chemco's perceived emphasis on lead time reduction programs, set-up time reduction, and flow management also seemed to be less than for the reference group. A benchmark study carried out six months earlier by a manager in the technology group had come to the same conclusions, but his colleagues had never accepted the results of his analysis. The fact that the Strategic Benchmarking Questionnaire suggested similar results led to a debate on the previous benchmarking study and a better analysis of its results.

5. Find the Questions

6. Search Out the Answers

The result of the debate in this first presentation and discussion was an agreement on the disagreement. The participants accepted that the issue of flexibility was far less clear than was assumed before the meeting and not very well internalized in the organization. But they did not necessarily accept the combined results of the previous benchmarking study and the questionnaire. In fact, the researcher who led the discussion suggested that the group should be careful with the results from the questionnaire. Strategic benchmarking is just a beginning point from which more detailed benchmarking efforts must emanate.

Three task forces of two people each were assigned to do three other functional benchmarking studies outside of the company, to evaluate the ongoing action programs in technology and systems, and to start thinking about a control system adapted to the needs of mix flexibility. The senior manufacturing manager also committed to writing a clear statement of what he expected to be the core capabilities needed in Chemco's manufacturing system. In order to do so, he involved the marketing division of Chemco who worked to develop a better understanding of customer needs.

7. Act On the Results and Start Over Again

About two months later, first reports by the three task forces were ready. These reports confirmed the original confusion about what mix flexibility means and what the resources and action plans were that would be needed to build a strong competitive position based on flexibility. The group decided to organize a three-day retreat with a representative from sales, marketing, and product development to further discuss the issue of mix flexibility. The retreat resulted in consensus on the role of flexibility in the strategy and operations of the company, changes in performance measures, and a comprehensive program of investments in technology, systems, and training of the workforce. The company also decided to continue the functional benchmarking started by the task forces and to discuss their reports and recommendations as more information evolved in the future. They had developed a new priority list of the key questions that they needed to broach in their functional benchmarking activities.

In an informal evaluation about one year later, the senior manufacturing manager told us that the most important result of the retreat was not the list of investments and actions, but the creation of a new reflex within his organization. Now, whenever a conflict occurred, or a major decision had to be taken, the organization referred back to the strategic benchmarking efforts and their shared assumptions about strategic intent and competitive status.

EXAMPLE 2: AUTOCO

Autoco is an American automobile producer that had been under attack from foreign competition for some time. It had responded positively with a number of improvement efforts, including TQM (total quality management), JIT (just-in-time), automation, functional benchmarking, worker participation, and so on. The problem in one large division of this company was that it had become lost in its own improvement efforts. There were dozens of different campaigns for improvement, but none seemed to be having a result, all were competing for attention and resources, and none seemed to tie together in a way that made sense. Both management and the rank and file were discouraged and frustrated—they wanted to improve, but until everyone started rowing the boat in the same direction and differences were resolved about what improvement meant and how it was to be achieved, it was difficult to see how they could succeed. Management decided to strategically benchmark the division in order to get everyone back to the same starting point and speaking the same language as they thought about the manufacturing strategy of the future and the specific practices that should be benchmarked.

1. Fill Out the Strategic Benchmarking Questionnaire

Sixty-one members of the division representing both the union and management from three plants and division headquarters participated in a strategic benchmarking workshop. Table 4–1 shows the breakdown of personnel from each part of the organization by area of specialization. Workshop participants were identified by a team of senior managers representing the company and the union to ensure a representative cross-section of employees. The workshop was introduced by the division manager, after which the participants were asked to fill out the Strategic Benchmarking Questionnaire in use at that time.

2. Benchmark (compare) the Questionnaire Results Internally

The individual responses to the questionnaire were entered into a computer database for intensive analysis focused on the differences in perceptions and beliefs between plants,

TABLE 4–1
Participants by Plant and Function

	Division	Plant 1	Plant 2	Plant 3	Total
Union	5	4	6	4	19
Quality	1	0	2	2	5
Management	7	2	4	4	17
Personnel/Human resources	3	1	1	1	6
Production engineering	0	3	0	2	5
Finance	1	1	1	0	3
Other	2	3	1	0	6
Total	19	14	15	13	61

functions, and levels in the division. This analysis points out the extent and nature of communications problems, or other differences.

Statistical analyses were prepared from the responses of the 61 individuals who returned their completed survey instruments. Three separate analyses of the differences in responses were made (1) between plants, (2) between levels in the organization, and (3) between groups with different functional responsibilities. While the three sets of groups differed, the same basic procedures were followed in each analysis.

First, the data were prepared for analysis by computing the difference between each individual's response on a survey item and the mean for all individuals on that item. This procedure isolates the differences in response within each section of the survey.

Each question in the questionnaire instrument was also separately analyzed with standard statistical procedures.[1] Some of the conclusions developed as a result of this analysis are indicated as follows.

Differences between Plants. Significant differences existed between plants on the importance of quality, delivery speed, and delivery reliability capabilities. Plant 1 emphasized these three capabilities to the greatest extent. Plant 2 rated quality significantly lower, while Plant 3 rated delivery speed and delivery reliability as less important than the other plants. Did these differences represent legitimate characteristics of the charters assigned to each plant, or was there confusion about

---

[1]The statistics for each of these analyses were derived from an analysis of variance (ANOVA) performed on each data set. Key statistics include the p-value, the group means, and the Fisher and Scheffe test results. The p-value, indicates the level of significance of the ANOVA. We used a cutoff value of .15, which means that we considered any p-value of .15 or less as indicating a significant difference in responses across the groups taken together. The group mean values indicate the group mean of the transformed data set. The Fisher PLSD and Scheffe F-test indicate whether significant differences exist between individual pairs of groups, or across all groups.

what was important? These are but two of the many specific questions which confronted the company.

Differences between Levels. An analysis of differences similar to that carried out between plants was also done by level in the organization. The three levels compared included union employees, plant level non-union employees, and divisional level employees. Significant differences appeared in the level analysis, particularly between division employees and the union, although some interesting differences between divisional and plant level employees also appeared.

Divisional employees rated the importance of quality, product performance, volume flexibility, and flexibility to introduce new products and design changes significantly higher than union representatives. Plant level employees also rated quality and product performance more important than the union. However, plant level employees joined with the union in judging flexibility to introduce new products and design changes as less important than did division employees.

Interestingly, both union and plant employees felt that the organization was significantly weaker in product and design change flexibility than divisional employees. There was general agreement that the organization's capability to compete on conformance quality was weak, but the union felt there was a much greater weakness in product performance than did division and plant personnel.

There was a great deal of disagreement about the past effectiveness of various actions to improve the organization's capabilities. The division saw changes in process technologies, sourcing patterns, organization, information systems, and quality procedures as much more effective. The plants saw changes in planning, procurement, and performance measurement procedures as very effective and joined with division employees in their assessment of organization changes and quality procedures as effective. Union employees rated most items as significantly less effective, though they joined with division employees in rating changes in information systems as highly effective, while the plants did not. Divisional and union employees also disagreed with plant employees in their assessment of the effectiveness of improved planning procedures. The plants rated these changes as significantly more effective than either the division or the union.

Differences between Function. An analysis of the differences in responses between six functions was also performed. The six functions included union, management (both plant and divisional), quality, finance, production engineering, and HR (human resources/personnel). Due to the small sample size of several of these groups, this analysis was least conclusive. For the most part, the results mirrored those identified by an examination of differences in levels—in particular, the differences in perception between plant and divisional employees and the union. Therefore, we will highlight here only those significant differences between various functions at the plant and divisional levels.

Production engineering and HR/personnel employees exhibited the greatest differences in their assessments of the effectiveness of past and future actions for improvement. HR/personnel judged past changes in performance measures and rewards to be less effective in reinforcing the development of key capabilities, disagreeing with their counterparts in management, finance, and production engineering. Production engineering sided with union employees in rating the past effectiveness of changes in organization and quality procedures less effective than their counterparts in management, personnel, and financial functions.

In all three analyses of difference by plant, level, and function, we see evidence of the confusion and lack of consensus within the organization, which of course led to this benchmarking project in the first place. However, by providing specific documentation about where, specifically, there is agreement and disagreement, the organization will be able to challenge assumptions, and develop a path for productive resolution.

3. Benchmark the Results Against Those of Other Reference Groups

Overall mean responses were developed and compared to the results of the benchmark reference group that was selected for analysis. The company decided to compare itself to all U.S. and Japanese manufacturers. The comparisons were made to aggregate data similar to that found in the tables in Part 3 of this book. This analysis indicated how the company's perceptions were different or similar to those of a group of leading U.S. manufacturers.

Table 4–2 shows the mean scores for the responses of the workshop group compared to the mean responses on competitive capabilities for the 1988 U.S. and Japanese Manufacturing Futures Survey.

4. Analyze the Differences

The top set of data in Table 4–2 shows the mean importance score given to each competitive capability by the average U.S. and Japanese firm in the Manufacturing Futures Survey sample and the mean for the Autoco group. The scale used to develop the scores ranged from 1 (low importance) to 7 (high importance). The asterisks (*) identify the capabilities for which there was a statistically significant difference between Autoco and the reference group; that is, for which there is less than a 5 percent probability that the differences were due to chance. This comparison suggests that Autoco employees see low prices as more important than the reference group, and service as less important. In comparison to the Japanese, low prices and new product flexibility were seen as much less important.

The bottom set of data in Table 4–2 indicates the mean strength scores of Autoco and the reference groups. A seven-point scale, where 1 corresponds to weak and 7 to strong, is used to indicate respondent *perceptions* about their organization's current competitive strength on each capability, relative

TABLE 4–2
Capability Assessment: Autoco versus U.S. and Japanese Norms

	U.S. Mean	Japanese Mean	Autoco Mean
Capability Importance			
To offer low prices	5.26	6.50	5.74*
To change design and introduce new products	5.60	6.30	5.75
To make rapid volume changes	4.84	5.90	4.82
To offer consistent quality	6.54	6.30	6.56
To offer high-performance products	5.79	6.30	5.64
To provide fast deliveries	5.67	6.00	5.54
To make dependable delivery promises	6.15	6.00	6.13
To offer after-sales service	5.39	5.60	4.80*
To offer a broad line	5.01	5.20	4.90
Strength on Capability			
To offer low prices	4.18	4.40	4.07
To change design and introduce new products	4.49	4.60	3.31*
To make rapid volume changes	4.68	4.60	5.03*
To offer consistent quality	5.53	5.20	4.23*
To offer high-performance products	5.28	5.10	5.30
To provide fast deliveries	4.87	4.50	5.15
To make dependable delivery promises	5.02	4.60	5.05
To offer after-sales service	5.04	4.60	4.37*
To offer a broad line	4.93	4.40	5.38*

*Difference between Autoco and reference group is statistically significant.

to the competitor who does it best in their industry. The data suggest that Autoco employees see themselves as stronger than the U.S. reference group on delivery speed, in having a broad product line, and volume flexibility. The respondents interpreted this to mean that they believe they have a relatively strong ability to respond to a delivery crisis quickly. But they also saw themselves as weaker than the reference groups on conformance quality, price, service, and the ability to quickly introduce new products or make changes in design.

5. Find the Questions

In order to develop more consensus about where they stood and where they wanted to go, the group was then split up into four teams representing the cross-section of employees from each of the three plants and division HQ. The teams were asked to analyze and discuss their responses and to prepare an oral report on their findings to all workshop participants. Without bringing more information into the discussion from outside, it was impossible to bring all of the issues to a conclusion. However, this process served to develop consensus on the priorities for the various competitive capabilities and the broad areas in which further benchmarking, analysis, and action were most likely to yield improvement.

All of the groups came to agree that quality should be their first priority and that a new quality program that was being proposed should become the engine of change that would drive their common efforts. Beyond specific actions for improvement in quality, the groups also identified the potential they saw for improving by changing their policies or methods in a number of broad areas relevant to manufacturing strategy. Table 4–3 shows what the statistical analysis indicated about the consensus view on the most fruitful areas for future work.

The third column in Table 4–3 indicates the mean score given each action area in terms of its future potential to contribute to improvements in critical capabilities. (Items are scored from 1, no potential, to 7, great potential). Nearly every item has a higher score, compared to the second column. The asterisks (*) indicate the items that are judged to have the greatest impact compared to the historical level of reinforcement. Overall, further changes in quality procedures and changes in human resources management are judged to have the greatest

TABLE 4–3
Historic and Future Manufacturing Strategies

Action Area	Reinforcement Last Two Years	Potential Next Two Years
Capacity	4.44	4.37
Facilities	4.53	4.80
Sourcing	4.07	5.25*
Process technology	4.68	5.89*
Information systems	3.95	5.38*
Human resources policies	4.72	6.21*
Performance measures and rewards	3.69	5.23*
Organization	4.41	5.16*
Quality procedures	5.02	6.26*
Product/process development procedures	4.39	5.64*
Planning procedures	4.18	5.39*
Procurement methods	4.05	4.83

*Items with greatest impact compared to historical level of reinforcement.

potential for improvement. The difference in scores between columns 2 and 3 shows that the greatest untapped potential lies in changes in human resources, information systems, and performance measures and rewards.

6. Search Out the Answers

7. Act On the Results

It is important to note that the first actions taken as a result of the strategic benchmarking workshop were not to go out and do more benchmarking or to gather other information. Rather, the group decided that the first thing they would do was to *act* on the key priorities they had agreed on. They realized that their lack of understanding and agreement had been the biggest barrier to action, and that they knew much of what needed to be done to improve rapidly. Several very specific actions for improving quality were identified for immediate implementation.

It is also interesting to note that the first place they recommended seeking more information was not externally but internally. Specifically, they saw that they needed to think through the differences between their plants and how each could maximize its contribution to the competitive health of the division. This action was a part of a larger effort to develop a manufacturing strategy.

EXAMPLE 3: DIVERSICO

Diversico Products Company is a major manufacturer of containers and packaging materials to consumer and industrial markets worldwide. Until the late 1970s, the planning process at Diversico had a predominantly financial orientation; the early 1980s were characterized by the addition of marketing-related planning concepts. PIMS (profit impact of marketing strategy) data have been used in combination with other techniques to assess shareholder value and market attractiveness. Perceived quality surveys (PQS) were employed to assess customer needs and competitive position. As the company looked forward to doing business in the 1990s, it began to further develop the planning process to address additional linkages—those that would bring finance and marketing together with manufacturing. Diversico was a manufacturing company, and it became obvious that operational issues, technology, and human resources planning had to be better integrated into the planning process.

The company began to search for methods to deal with its new planning needs. Several outside consultants were brought in to educate Diversico management, along with researchers from the Manufacturing Futures Project. After working through this initial education period, the company decided to implement a pilot project. The paper division became the first to attempt development of a comprehensive manufacturing strategy, using the capability development methods from the survey to establish the needs and views of their customers and suppliers as a first step. They accomplished this by scheduling a series of face-to-face meetings between top-management groups in which each discussed views on various capabilities, and in particular, what was meant by them.

A year later, the paper division presented its progress to top management. Results of this work were encouraging. The division began to articulate a detailed plant focus and quality strategy that addressed business capacity needs and at the same time facilitated improvements in quality capabilities. An important outcome of the process of using the questionnaire was an improved relationship between the division and its customers. A great deal of time was spent with divisional customers, listening to them explain the capabilities they most valued and how the paper division could support them. Better definitions of product performance evolved, as well as an understanding of how to focus the paper division strategy to consumer needs.

This positive development supported the next step of the process—to initiate a companywide implementation of a manufacturing strategy process. However, this was not considered an easy task, as Diversico operates a broad set of increasingly diversified operations worldwide. Moreover, it was assumed that it would take time to develop the necessary skills and processes across the organization. Therefore, a three-phase approach to manufacturing strategy planning was planned:

1. Create manufacturing strategy awareness.
2. Develop business unit manufacturing plans.
3. Improve linkages to other strategic functions.

Phase 1: Create Manufacturing Strategy Awareness

As Diversico began its annual planning cycle for the next year, it initiated a companywide program that sought to create a broad awareness of manufacturing strategy and consensus in key capabilities. In order to accomplish this, the company used the 1988 Manufacturing Futures Survey to do strategic benchmarking. Administered to the company's top manufacturing managers in 17 business units worldwide, the survey proved its worth by focusing attention on manufacturing strategy and drawing together facts and perceptions about Diversico manufacturing. Lessons learned from this exercise include the following:

1. The survey helped the divisions to assess their abilities and to prioritize the importance of developing certain critical capabilities. This provided a framework for initiating the strategy formulation process and focused attention on particular strengths and weaknesses. Furthermore, it helped develop consensus about the few capabilities that the corporation stressed in all its divisions.

2. The strategic benchmarking survey results provided an assessment of divisional action programs that were currently in place or perceived to be important. It raised questions about how these programs would or could improve key capabilities, and how they were related to one another.

3. The survey proved to be a fairly "painless" way of developing a common definition of manufacturing strategy in the corporation. It required a relatively small investment in time, and yielded a provocative, yet nonthreatening, evaluation of manufacturing in each division. The survey was even more effective when divisional teams of managers were included in its development. When interfunctional teams of managers were involved in answering the survey, the most important product of the effort was debate and the subsequent increase in understanding of key manufacturing issues.

Phase 2: Develop Business Unit Manufacturing Plans

In the next planning year, the second phase of the program began. This phase of the planning process was intended to push both manufacturing and general managers beyond the consideration of generic capabilities. The business unit manufacturing managers were requested to speak to specific business unit issues and plans and to articulate manufacturing strategies that addressed these situations. The following are lessons learned from this exercise:

1. Involving manufacturing executives directly in the planning process served to encourage their personal commitment to the strategic process.

2. Effective manufacturing planning requires hard measures of strategic performance and factual benchmarks. Without these, real competitive analysis and progress toward strategic goals cannot be monitored.

3. Functional planning requires a "dedicated" forum. Manufacturing-specific concerns cannot be adequately addressed in a generalized strategic forum that covers all functions. Separate functional reviews are necessary to provide the depth.

4. Again, the benefits of a cross-functional team approach to manufacturing strategic benchmarking and strategy development was found to be rewarding. It usually led to better integrated and articulated plans.

Phase 3: Improve Linkages to Other Strategic Functions

As the company enters the 1990s, it is moving ahead to emphasize the role of manufacturing within the multifunctional environment of business strategy work, and improving the linkages between business plan development and capability development. The notion underlying all this activity is that business strategy is ultimately a general management responsibility in which all functions, and all levels of employees, are playing in concert.

Currently, attention is being focused on the interaction between manufacturing and marketing, and in particular, on the language and concepts they employ. At present, for example, there are several "dialects" being used to discuss quality. Marketing focuses on "perceived" quality, while manufacturing is focused on "conformance" quality. Both concepts are important to understand in order to translate customer needs and perceptions into meaningful manufacturing capabilities. Diversico is now seeking to establish competitive capabilities and the total quality process as the structure for developing a new common language for its perceived quality surveys. When this is accomplished, both marketing and manufacturing can use the same language to address priorities and goals.

PART 2

MANUFACTURING AROUND THE GLOBE

CHAPTER 5

THE UNITED STATES: BEYOND THE QUALITY REVOLUTION*

During the 1980s, quality was the driving theme in U.S. manufacturing. This theme overtook American industry so suddenly and completely that, in previous Manufacturing Futures Reports, we have labeled the transformation the *quality revolution*. Today, quality is still a dominant theme in American industry; and, unless manufacturers forget the lessons of the past, it will always be so. However, significant changes are also signaled by large successful U.S. manufacturers that are moving beyond the quality revolution to address ambitious plans for growth in overseas markets, price competition, and the need for faster product development. Competitiveness is only fleetingly in equilibrium. As soon as industry becomes competitive on one dimension (such as quality), competitive forces and world events raise new standards to be met by other dimensions.

How competitive are U.S. manufacturers? Have they improved their competitive position, and if so, how? Looking back at the results of the combined 1988 Manufacturing Futures Survey of the United States, Japan, and Europe, it is apparent that the leading U.S. manufacturers in our sample believe they have gained on a number of dimensions of performance. Figure 5–1

---

*This chapter has been adapted from "Beyond the Quality Revolution" by Professors Jeffrey G. Miller and Jay Kim, Boston University Manufacturing Roundtable, 1990.

FIGURE 5–1
Annual Rate of Performance Improvement, 1985–1987

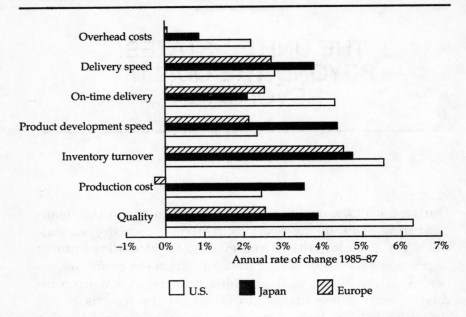

depicts the relative rates of improvement on key competitive variables from 1985 to 1987. It shows that the U.S. firms in the sample improved their performance in quality, inventory turnover, manufacturing overhead cost, and delivery dependability at a faster rate than a comparable set of competitors in other critical countries. The Japanese showed a faster rate of improvement in their ability to introduce new products, reduce delivery lead times, and reduce costs, while the Europeans lagged in all of the dimensions of performance measured.

The 1990 U.S. survey shows that the U.S. firms in the sample continued to improve their quality and delivery dependability over the 1988–89 period at about the same rate as in 1985–87 (see Figure 5–2). The annual rate of improvement in inventory turnover increased significantly from 5.5 percent to 11.5 percent; the rates of improvement in cost and product development time increased somewhat. During the period from 1985 to 1987, U.S. unit costs were reduced at an annual rate of 2.5

FIGURE 5–2
Annual Rate of U.S. Performance Improvement, 1987–1990

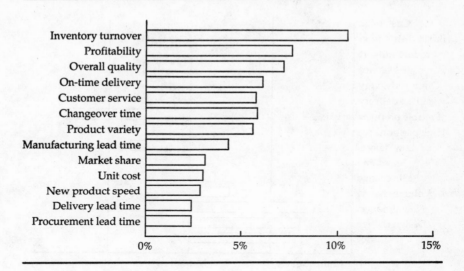

percent per year, and product development time was reduced by 2.1 percent per year. From the beginning of 1988 through the end of 1989, the annual improvement rates in cost and development time were 2.9 percent and 2.8 percent, respectively. Other dimensions showing significant improvements include profitability, customer service, changeover or set-up time, and greater product variety from existing factories.

The competitive position of large U.S. firms has improved in areas where they have enjoyed a faster rate of improvement, but rates of improvement do not tell us about absolute competitive position. Figure 5–3 shows how the sample of U.S. firms perceive their competitive position with respect to their most important competitors (domestic or foreign) on various dimensions. Over 80 percent of the respondents indicated competitive strength in these areas: reliable products, **high-performance product*** design, and low defect rates. The U.S. firms also do

---

*Terms defined in the glossary at the end of Part 3 are in boldface.

FIGURE 5–3
Relative Competitiveness, 1990

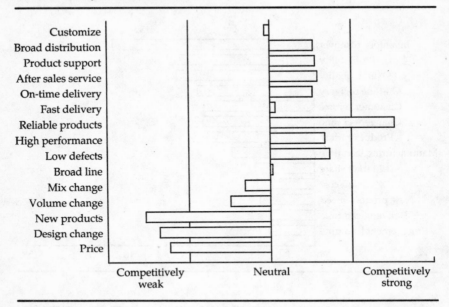

well on a number of service dimensions such as broad distribution, **product support, after-sales service,** and on-time delivery.

The major competitive weaknesses are in flexibility and the ability to compete in price competitive markets. Over 50 percent of the respondents indicated competitive weakness in their ability to develop or change products quickly. Nearly half of the respondents indicated weakness in their ability to compete on price. We are concerned to see that the areas of greatest competitive weakness are those in which the slowest progress has been made.

EFFECTIVE MANUFACTURING STRATEGIES IN THE 1980s

The performance improvements described in the previous section can be attributed to a variety of actions that were taken by manufacturing companies in the 1980s. Table 5–1 shows the

TABLE 5–1
Changes in Manufacturing Strategy Portfolios

Top Five Action Plans, 1984	Top Five Action Plans, 1986	Top Five Action Plans, 1988
Production inventory systems	SQC	Vendor quality
Reduction in workforce	Zero defects	SQC
Supervisor training	Vendor quality	Worker safety
Direct labor motivation	Improving new product introduction capability	Manufacturing strategy
Process/product development	Production inventory systems	Worker training

top five most popular action plans in the manufacturing strategy portfolio of the typical U.S. senior manufacturing executives who responded to the Manufacturing Futures Survey from 1984 through 1988. The table reveals an interesting series of transitions, starting with a preoccupation with production inventory systems and downsizing in the mid 80s, to the beginning of the quality revolution in 1986, to more emphasis on supplier quality and manufacturing strategy in 1988.

The table also highlights a notable shift in the attitude of manufacturers toward the role of production workers. In the mid-80s, supervisor training was emphasized, and direct labor was perceived as something to be simply motivated. In the late 1980s, we see that attention and training shifted to the worker, a figure now seen as playing a larger role in the factory.

Respondents to the 1990 survey were asked to indicate the relative pay-offs that had resulted from the actions listed in Table 5–1, as well as other activities, over the period 1988–89. The top five actions with the greatest pay-offs are shown in Table 5–2, while Table 5–3 indicates the actions that had the least pay-off as of 1988–89. Note that some of the most effective actions were not included in Table 5–1.

Over 58 percent of the respondents had used inter-functional work teams during 1988–89, and felt that this was the most effective activity in which they had engaged. These

TABLE 5–2
Most Effective Actions, 1988–1989

- Interfunctional work teams (58 percent)
- Manufacturing reorganization (58 percent)
- SQC (68 percent)
- Manufacturing strategy (70 percent)
- Just-in-time (51 percent)

Note: Percentages are of respondent firms that adopted each particular action in the 1988–89 time frame.

TABLE 5–3
Least Effective Actions, 1988–1989

- Robotics (18 percent)
- Activity-based costing (22 percent)
- Integrating information systems in manufacturing (58 percent)
- Integrating information systems across functions (48 percent)
- Flexible manufacturing systems (30 percent)

Note: Percentages are of firms that adopted each particular action in the 1988–89 time frame.

interfunctional teams participated in a variety of activities ranging from quality improvement to reductions in production lead time, time to develop new products, and changeover time. The power of these teams to cut across traditional functional boundaries to attack systemwide problems is apparent from the high pay-off attributed to them.

Curiously, manufacturing reorganization, which had received low marks for effectiveness in the 1988 survey, received high marks in 1990. This change may suggest that it takes a long period of time for some actions to generate tangible pay-offs. Closer working relationships developed in the process of building interfunctional teams, the emergence of more flexible attitudes toward organizational change, and better manufacturing strategies may also explain why organization change is viewed more positively.

SQC (statistical quality control), with over 68 percent of the respondents emphasizing it in the past two years, was rated highly, justifying its persistence in the action portfolios of manufacturers over the 1986–90 time frame. The high pay-off activity in which the most respondents (70 percent) were engaged was the development of a manufacturing strategy that linked manufacturing activity to the overall business strategy. Just-in-time methods have also had a high impact for the 51 percent of the sample that have employed them.

The list of least effective actions must be approached carefully. It does not indicate that these actions had no value, only that they had relatively low two-year payouts in comparison to the 26 other action items presented in the survey. Typically, many of these action programs have relatively long payback periods. However, it is notable that most of the activities listed in Table 5–3 are concerned with the adoption of new technology. This suggests that the time required to implement new technologies is still much too long. Nevertheless, the high participation rates for integrating both manufacturing and interfunctional information systems show that manufacturing firms are not shying away from these longer term investments.

BUSINESS STRATEGIES FOR THE 1990s

Looking ahead into the 1990s, the survey shows that the respondents are planning to take an aggressive posture. In 1990, over 79 percent of the respondents indicated that their business strategies were focused on increasing market share (see Table 5–4).

We offer three main reasons for this offensive posture. First, U.S. manufacturers are in a better competitive position with respect to quality, as noted earlier. Second, U.S. manufacturers left the 1980s with several years of record profits, positioning them to invest in their businesses to a greater degree, despite the recession in the early 1990s. Third, the fall of the

TABLE 5–4
Market Strategies for 1990

	Percent of Respondents
• Build market share	79%
• Hold (defend) market share	18
• Harvest	2
• Withdraw	1
	100%

iron curtain and the unification of Europe are creating major opportunities internationally.

The aggressiveness of U.S. business strategies can be seen more clearly by examining where they intend to grow. Figure 5–4 shows the proportion of offshore sales, production, and sourcing that the respondents are planning as a percent of total activity for 1992 compared to 1989. Offshore sales are planned to grow to 26 percent of total sales in 1992 from 22 percent in 1989. This implies that offshore sales will grow at more than twice the rate of domestic sales during this period. Offshore production and sourcing are also planned to grow, but at a

FIGURE 5–4
Increased Offshore Activities

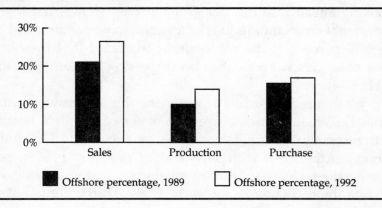

Offshore percentage, 1989   Offshore percentage, 1992

slower pace than offshore sales. Clearly, the main targets of aggressive activity are outside the U.S. market.

THE CHANGING FACE OF COMPETITIVENESS

The desire to build market share and to expand internationally is changing the face of competitiveness. For some time, competitiveness and quality have been viewed almost synonymously in the United States. More recently, we have seen more attention given to time-based competition; that is, emphasizing delivery dependability, and reducing lead times in new product development, production, and procurement.

Conformance quality and dependable delivery are still numbers one and two in the competitive priorities of businesses in 1990, followed by performance quality, price, and fast delivery in order of importance. However, there has been a substantial shift in thinking in 1990 compared to 1988. Figure 5–5 highlights the significant increase in importance of broad distribution, a broad product line, and price as key competitive capabilities. These changes are consistent with the intended

FIGURE 5–5
Competitive Capabilities: Percent Changes in Relative Importance, 1988–1990

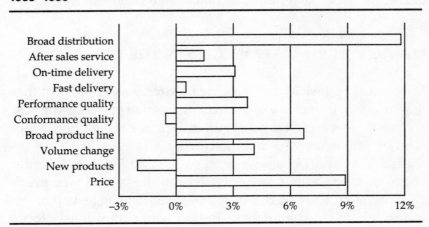

expansion of U.S. manufacturers into global markets. A global customer base requires the development of a more extensive factory-distribution network. Catering to local tastes around the world requires broader product lines and factories geared up to produce them. Better quality and delivery performance across an industry means that competitors find it more difficult to differentiate themselves; the result is that price becomes a relatively more important factor in competition.

Curiously, the importance of rapid new product development did not change significantly between 1988 and 1990; if anything, it was reduced in importance. It remained the sixth most important competitive priority in the typical business strategy, even though the respondents see this as an area of significant competitive weakness (see Figure 5–3). We believe that this indicates that vulnerability to rapid new competitive product introductions is not perceived as a highly serious issue. But the international Manufacturing Futures Surveys have shown that the Japanese, in particular, intend to exploit this weakness (see Chapter 7). Moreover, conformance quality remains the highest ranked in importance, even though it is widely accepted that it is becoming more difficult to differentiate oneself from the competition on this basis. We are concerned that the nationwide emphasis on conformance quality, and the previous success of these firms in improving it, may be distorting their perception of competitive realities.

MANUFACTURING STRATEGIES IN THE 1990s

Data from the 1990 Manufacturing Futures Survey suggest that the influence of recent experience, aggressive plans for the 90s, and shifts in perceived competitive factors have all had a significant impact on the manufacturing strategies that are intended to support business strategies. Table 5–5 lists in order of importance the five key objectives of the typical vice president of manufacturing through 1992. This list suggests that the future role of the manufacturing unit is to continuously focus

TABLE 5–5
Manufacturing's Role: Top Five Manufacturing
Objectives, 1990

1. Improve conformance quality
2. Improve vendor quality
3. Reduce unit costs
4. Reduce overhead costs
5. Reduce product development cycle

on quality in the supply pipeline, reduce unit costs (particularly overhead costs), and shorten the time it takes to develop and roll out new products.

The first two objectives are consistent with the continued strategic importance that manufacturing firms are placing on quality as a means of maintaining market position and reducing the costs of nonconformance. The emphasis on cost, the third and fourth objectives, is consistent with the increasing importance of price competition. Figure 5–6 shows the percentage reductions anticipated in the main elements of manufacturing cost by 1992. The figure indicates that direct labor and overhead costs will play a key role in cost-reduction efforts.

The inconsistency we see is the lack of emphasis given to on-time delivery, which was the second highest ranked business

FIGURE 5–6
Predicted Reductions in Manufacturing Cost Elements, 1989–1992

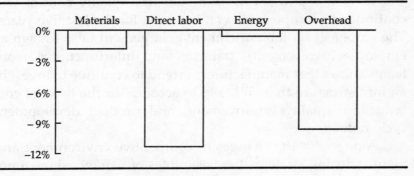

TABLE 5–6
Top Five Action Plans, 1990

1. Linking manufacturing and business strategies
2. Job enlargement and enrichment
3. Statistical quality control
4. Worker and supervisor training
5. Interfunctional work teams

priority, and the greater emphasis given to reducing the product development cycle, despite its lower ranking as a business priority. The anomaly implies differences in the views of manufacturing executives and their superiors in the business unit.

The actions that manufacturers plan to take to achieve their key objectives in quality, cost, and new product speed vary widely by company and industry. However, there is a common thread, and Table 5–6 lists the top five action plans of manufacturing managers during 1990–92. The table shows that linking manufacturing and business strategies is the top-rated activity. The rise in importance of this activity to U.S. manufacturers has been apparent for the last several years of the survey. Its popularity can be attributed to its importance in planning the structure of new factory-distribution networks and systems, which we assume manufacturers will require to achieve broader distribution, and its importance as a tool for resolving the apparent inconsistencies we have noticed between business strategies and manufacturing activities.

Table 5–6 also shows that statistical quality control will continue to be important, as it has been for the last five years. The emphasis on job enrichment/enlargement (also known as employee involvement), training, and interfunctional work teams shows that manufacturers intend to continue to invest in an infrastructure that will help to accomplish the tasks of cost reduction, quality improvement, and product development cycle reductions.

Along with the changes in competitive environment and manufacturing strategy, the basic roles of workers show a no-

FIGURE 5–7
Percent of Direct Labor Time Spent on Indirect Activities

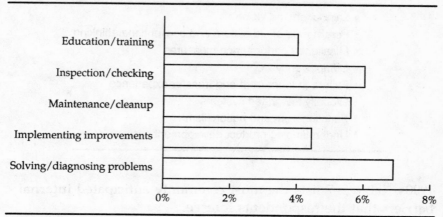

table trend. Figure 5–7 indicates the substantial amount of time that is spent by the average direct worker in the manufacturing firms on non-production activities. Nearly a third of all **direct labor** hours are spent in training, checking, maintaining, solving problems, and implementing improvements. Action plans that include a strong element of worker/supervisor training are necessary to ensure higher productivity in both direct and indirect tasks.

In sum, manufacturing strategies for the 1990s are generally consistent with the basic direction of the business units in the survey, though the inconsistency in addressing product development speed is troublesome. In comparison with the strategies revealed in previous years (see Table 5–1), the strategies for 1990–92 show a trend toward pushing the whole organization to work smarter, not just harder.

BARRIERS TO IMPLEMENTATION

The manufacturing executives that responded to the survey know that they will face a number of internal and external obstacles in implementing their manufacturing strategies in the

TABLE 5–7
Internal Barriers for the 1990s

- Lack of shared vision
- Persistence of old culture and conventional thinking
- Misaligned performance measures
- Complacency
- Failure to learn from and transfer experience
- Globally integrated systems
- Functionalism and nationalism
- Understanding product development cycle

1990s. Table 5–7 lists the most commonly anticipated internal barriers that the respondents foresee.

A shared vision of the future is necessary in order to direct, motivate, and explain needed changes, as well as to overcome the inertia of an established culture. Recognition of the importance of a shared vision and the inconsistencies we have observed with respect to delivery and product development speed explain the emphasis that manufacturing executives are placing on the development of a manufacturing strategy linked to the business strategy.

Over the last several years, our surveys have shown a significant change in the way that manufacturing companies communicate information about their vision and business strategies. As shown in Figure 5–8, there has been a marked trend toward sharing this information with all employees, rather than keeping it solely in the domain of top management. The 1990 survey shows a continuation of this trend.

In order to be effective, performance measures must be realigned with the strategy and action programs that are underway. We also see that complacency looms as a serious problem. Having made considerable progress on a number of performance dimensions during 1985–89, particularly with respect to quality, managers are finding it difficult to transmit to stakeholders the message that the job is not done.

FIGURE 5–8
Percent of Respondents Who Share Knowledge about Goals and Plans Widely and Narrowly

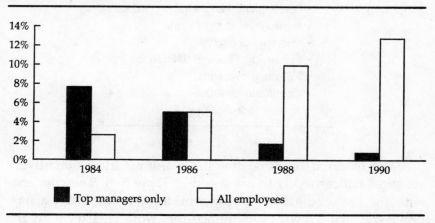

The last four internal barriers are concerned with difficulties in communicating within a manufacturing company. The failure to examine the successes and failures of one plant and transfer these learnings to another is an old problem, but one that takes on added importance in the 1990s environment. As manufacturing firms address global markets with international plant and procurement flows, better communication systems and the language, cultural, and attitudinal problems inherent in communicating across functions and national borders become more important.

Functional isolation explains why manufacturing executives feel that more information is needed about the product development cycle as well. More than half of the responding business units start their process engineering activities after most of the product engineering is finished. This results in a longer new product development lead time and missed opportunities in rapidly changing markets. The data suggest that we are a long way from concurrent manufacturing and design engineering, and that the first step in achieving this is to understand the product development cycle better.

TABLE 5–8
External Barriers for the 1990s

- Poorly trained and educated workers
- Environmental regulations
- Shortage of capital
- International cultural differences
- Reluctant suppliers
- Conflicting alliances
- Access to technology

The external barriers that the manufacturing executives are most concerned with are listed in Table 5–8. The concern with the basic education of the workforce is no doubt a national concern. How can manufacturers work smarter with an illiterate workforce? Today's environmental problems are also at issue. Manufacturers are concerned that meeting environmental regulations will sap their energies from other problems, or place them at a disadvantage compared to competitors in unregulated countries. Potential shortages of capital and high interest rates may have an impact on much needed investment in new technology development. Communication with international suppliers, allies, technology providers, and partners is also a critical concern.

THE LEADERSHIP CHALLENGE

The single most overwhelming lesson managers have taken from their experience in the 1980s was that attention to quality pays. Clearly, U.S. manufacturing executives feel that the focus on quality has had a positive impact on their competitive position. Doubtless, this is one of the major reasons that quality continues to figure prominently in their portfolio of action plans for the future. The second most common lesson was that teamwork and employee involvement are critical to undertaking effective changes. The third common lesson was that manufacturing leaders needed to have a workable approach to raising the performance expectations of their organizations. Many of the respondents noted that focusing on customer requirements and competitive benchmarking were effective ways of accomplishing this. The fourth lesson was that good, fast product design with appropriate early manufacturing inputs was critical to improving quality, reducing costs, and reacting quickly to customer needs. Perhaps the inconsistency between manufacturing strategies and business strategies could be resolved if the rest of the corporation understood this lesson as well as manufacturing executives. Finally, manufacturing managers learned that intensive training is essential to achieving all of the above.

It is clear to us that one part of the leadership challenge for the future is learning how to more effectively apply the lessons of the 1980s to the 1990s. Manufacturers must hold their gains while maintaining and even increasing their rates of improvement. As one individual said, "The first round of continuous improvement is easy compared to the difficulty of sustaining the effort." Recall that complacency was identified as one of the critical internal barriers to achieving the strategies of the 1990s. Some of the success that these firms have had in their efforts to improve quality and effectiveness is becoming a problem. Another respondent noted, "We learned that we needed to drive for *continuous* improvement in cost, quality, and customer service, because when we turn around, we see that our competitors are right on our heels."

The focus on cost in the 1990s was also widely apparent in the comments of the respondents. One respondent noted, "We must do even more with less." The face of competitiveness is changing. The focus on reducing manufacturing costs while sustaining the drive for continuous improvements will require that manufacturing managers continue to exercise their leadership skills internally. They must also turn more of their attention to leading efforts that transcend functional, supplier, partner, and national boundaries.

BEYOND THE QUALITY REVOLUTION

The Manufacturing Futures Survey focuses on *what* leading manufacturers have accomplished and *how*, what they *believe* is important, and what they *plan* for the future. The questions that the survey does not directly address are: Are their beliefs correct? Are their plans well aimed? Does their understanding of what has worked well in the past provide a sufficient basis for actions in the future? We are certainly encouraged to see that significant progress has been made in many aspects of manufacturing competitiveness by these firms. Nevertheless, we are troubled by the areas where progress is not being made, by some inconsistencies in the findings and their possible implications, and by the barriers these executives face in the future.

The pace of improvement is slowest in the two areas where these companies are most behind: cost and product development speed. While maintaining the pace of improvement in quality, delivery, and inventory turnover, U.S. manufacturers must push themselves at a faster pace to address these weaknesses.

We are also concerned with the inconsistencies between the competitive priorities of business strategies and the key objectives of manufacturing executives in their manufacturing strategies. In particular, there appears to be a high need to carefully examine baseline assumptions about the importance of conformance quality and product development speed, and to develop a common vision of how the company will compete and how business and manufacturing strategies are linked. In particular, there seems to be too much emphasis placed on conformance quality, and too little recognition of the importance of good product development practices that provide quality through robust products and processes.

U.S. manufacturing managers must also be aware of the potential danger of becoming victims of their current successes, methods, and assumptions. The dominant activities in the portfolio of manufacturing strategies now includes developing manufacturing strategy and linking it with business strategies, facilitating employee involvement and development, and maintaining continuous improvements in quality, cost, and time. However, there are several key issues that this portfolio does not address. First, these companies have neglected a strong emphasis on developing new process technologies: They are certainly having difficulty in doing so with fast payouts. They should not forget the importance of learning how to develop technological potential while they concentrate on organizational and infrastructural development. Second, there is little evidence in the survey that these manufacturers plan significant structural change as they address new markets. Even though the need to develop stronger global factory-distribution systems is apparent, there is little in their future plans that addresses the actions necessary to support an aggressive international thrust. Finally, we believe that manufacturing managers and their superiors must take more proactive roles in overcoming various barriers. For example, they must consider approaches that capitalize on environmental concerns, rather than simply viewing the environment as a regulatory problem.

Manufacturing managers must avoid being trapped by the *conformance* quality/continuous improvement paradigm currently in vogue. They must develop and communicate to the entire organization a wider vision of the continuously changing face of the competition. They must lead U.S. manufacturing companies beyond the quality revolution of the 1980s that focused on defects, and move toward customer satisfaction.

CHAPTER 6

THE EUROPE REPORT: REMOVING THE BARRIERS IN MANUFACTURING*

As they enter the 1990s, many manufacturers in Europe find themselves in much more favorable circumstances than they did in entering the 1980s. This relative prosperity is due to generally favorable business prospects and the expectations created by the European single market. But this favorable climate also reflects the investment made by European manufacturing industry during the 1980s in cleaning up and revamping its factories. This reorganization of manufacturing and reconditioning of the plants was stimulated by a strategy focused on improving the quality of output and dependability of deliveries.

Efforts to improve total quality and other production capabilities are projected to continue. However, our sample of high-performing European manufacturers is also looking for new ideas, especially those that promise to improve the integration of the production function with the rest of the enterprise and its environment. A more systematic view of the logistics chain from supplier to final customer, more customer-driven manufacturing, and increased integration with other business functions have emerged as the new objectives of manufacturing

---

*This chapter has been adapted from Professors Arnoud De Meyer and Kasra Ferdows, INSEAD (European Institute of Business Administration), "The Europe Report: Removing the Barriers in Manufacturing," with their permission.

managers. This challenge is all the more daunting because of the turbulent changes currently underway in Europe. But for those who succeed, there is the promise of a highly competitive European factory.

EUROPEAN MANUFACTURING PERFORMANCE, 1985–1989

The European manufacturers represented in our sample have shown a remarkable recovery in the 1985–89 period. In 1983, when we started our survey, one out of three manufacturers reported a loss. In 1984, the ratio was 1 out of 4, and in 1985 only 1 out of 10 reported a loss. By the end of 1989, and despite predictions for a recession in the early 1990s, the picture looked very different from earlier years. For the 224 manufacturers in our 1990 sample, the average pre-tax return on assets for the last fiscal year was 17.8 percent, average unit growth rate was 13 percent, and capacity utilization was increased to 84.3 percent. These good results are surely due in part to better performance by manufacturing. Indeed, as shown in Table 6–1, most of the respondents report improvements on a number of specific manufacturing performance indicators.

Both the average performance improvements and the proportion of business units that have improved their performance on different indicators are remarkable; the improvement in profitability was spectacular. More than 80 percent of the respondents report an improvement in outgoing quality, which shows that the efforts in quality improvement have started to pay off. Other indicators have improved at a slower rate. In particular, delivery lead time, **procurement lead time,** unit production cost, and speed of new product development show the smallest rate of improvement in the list.

The quantitative evaluations of the improvements are backed up by other evidence. Out of a list of 15 possible competitive abilities related to price, flexibility, quality, deliv-

TABLE 6–1
Manufacturing Performance Indicators

Indicator	1985–1987 Average Improvement (percent)	1987–1989 Average Improvement (percent)	1987–1989 Respondents Reporting Improvement (percent)
Outgoing quality		15%	83%
Inventory turnover	13%	15	68
On-time delivery	8	15	64
Return on manufacturing assets		15	78
Customer service		14	78
Overall quality	9	13	76
Manufacturing lead time		13	67
Set-up times		12	63
Variety of products		12	60
Market share		11	62
Delivery lead time	8	10	57
Average unit cost	0	8	62
Speed of new product development	6	8	50
Procurement lead time		8	55
Profitability		28	72

ery, and service, the respondents considered their businesses to be stronger than their best competitor on ability to offer consistently low defect rates, to provide high-performance products, and to provide reliable durable products. They were, however, weaker on their ability to compete on price, to introduce new products quickly, and to make rapid design changes. In the rapidly changing political and trade situation in Europe, it is difficult to assess whom these companies consider their competition. However, given the presence of barriers to Japanese manufacturers in most of the nations of Western Europe, it is probable that their list of competitors is smaller and more localized than that of the Americans or the Japanese.

ACTION PROGRAMS THAT HAVE PAID OFF

The improved performance of manufacturing in Europe has resulted from the successful implementation of a number of action programs. In Table 6–2, we show the 10 action programs that received the most attention and the top 10

TABLE 6–2
Past Action Programs and Their Pay-offs

Programs Emphasized (1988–90)	Pay-Offs
Top 10	
Worker training	Manufacturing reorganization
Integration of information systems in manufacturing	Closing plants
Management training	Develop new processes for old products
Manufacturing organization	Reconditioning plants
Supervisor training	Develop new processes for new products
Quality function deployment	Quality function deployment
Link manufacturing to business strategy	Link manufacturing to business strategy
Integrating information systems across functions	Just-in-time
Giving workers a broader range of tasks/responsibilities	Interfunctional teamwork
Statistical quality control	Supervisor training
Bottom 5	
Value analysis/redesign	Giving workers a broader range of tasks
Closing plants	Production planning and inventory control system
Activity-based costing	Integrating information systems across functions
Design for manufacture	Quality circles
Robots	Activity-based costing

programs that had the greatest pay-offs. Also in Table 6–2 we show the bottom five programs in each category. The highest pay-offs have come from structural changes in manufacturing: reorganizing manufacturing, closing or relocating plants, reconditioning of the physical plant, and investing in new processes. The favorable economic environment of the late 80s has allowed Europeans to revamp their facilities and organization, with market forces providing an incentive to do so.

After the structural changes, we find a set of four programs also with high pay-offs. These programs focus on reaching out beyond the factory borders, such as **quality function deployment (QFD).** QFD is "a set of techniques for determining and communicating customer needs and translating them into product and service design specifications and manufacturing methods" that brings the customer's voice into the plant. Linking manufacturing to the overall business strategy, coupled with the development of interfunctional work teams, integrates manufacturing with the other company functions. And since the successful implementation of just-in-time (JIT) requires a close collaboration between a company and its suppliers, the high pay-off attributed to JIT investments hints at successful upstream integration.

The list of the five action programs that were implemented in the past and that have, relatively speaking, the lowest pay-off is a mixed bag. Some of them, like integrating information systems across functions, have been heavily emphasized only in the last two years—probably not enough time for a pay-off yet. Others, like **activity-based costing (ABC),** are still unknown. ABC is new to many of these companies and it requires basic changes in accounting practices within the company. Here, too, it may be too early to show a high pay-off, particularly if the required changes are found to be more difficult than expected. Still other action programs may have turned sour—among them, **quality circles,** which were very popular in the past.

EXTERNAL PRESSURES

One of the major issues that looms on the horizon of Europe's industrial environment is the creation of the single market by January 1993. How will the single European market influence manufacturing? In Figure 6–1, we show the expected changes for seven key issues. Product customization and the number of markets served by each factory will increase. Furthermore, the number of non-European competitors entering the EC (European Community) market is expected to grow. On the other hand, the number of suppliers will shrink as firms consolidate duplicate facilities. The single market seems to have only a minor impact on the number of factories, the number of models produced per factory, or the organizational autonomy of factories. In other words, our respondents expect that their factory network will be simplified: the same number of factories, but

FIGURE 6–1
Expected Influence on 1992 Production

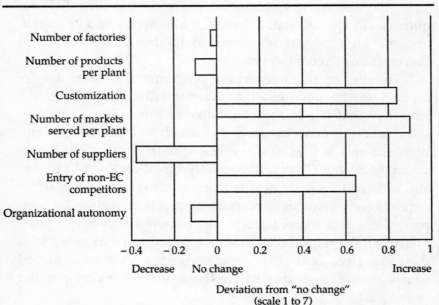

more markets served by each factory without increasing the product range, and with fewer suppliers. But this can only happen if these factories specialize in a few products, which they produce for the whole of Europe, as opposed to a single national market. And this will only happen, as Figure 6–1 shows, if the degree of customization with a reduced number of product families is increased.

We also asked our respondents to rate themselves against their best competitor in their ability to predict product, process, and market changes. European manufacturers single out one issue where they think they have more difficulties than their best competitors: the ability to cope with faster changes in products. However, they think they have fewer difficulties than their competitors in envisioning new processes and in their ability to attract capital resources for investment. This is partly explained by the sample: large manufacturers have less difficulty in obtaining capital resources and have enough in-house know-how to estimate their process capabilities.

Many of our respondents, though located in Europe, operate globally. Do they, and the rest of the sample, see a shift in production, sales, and procurement in 1992? As shown in Table 6–3, no big changes are expected in the short term in the geographical distribution of production and purchasing patterns. Sales patterns are expected to change significantly, however. The change from 17 percent exports to 20 percent exports in a

TABLE 6–3
Source of Production, Purchases, and Sales

	Production (percent)		Sales (percent)		Purchases (percent)	
	1989	1992	1989	1992	1989	1992
Europe	94%	93%	83%	80%	83%	83%
U.S./Canada	3	4	8	9	10	9
Asia Pacific	2	2	5	6	5	6
Others	1	1	4	5	2	2

two-year period represents a net export growth rate of about 9 percent per year.

NEXT STEP: REMOVE BORDERS

European manufacturers were able to build up a small war chest from the prosperity of the late 1980s. Their factories have been cleaned up and reorganized. What's next? For starters, more than two thirds of the respondents intend to build market share in the period from 1990 to 1995. This is a clear departure from our previous survey results in which respondents indicated the most important strategic direction was defending market share.

To be effective, this aggressive marketing stance must be reflected in the continuous upgrading of manufacturing capabilities. The rank ordering of the importance of 14 different capabilities shown in Table 6-4 indicates the general direction of these efforts for European manufacturers as a group in 1990.

TABLE 6-4
Rank Order of Competitive Priorities (First Is Most Important)

1. Offer consistently low defect rates
2. Offer dependable delivery promises
3. Provide reliable/durable products
4. Provide high-performance products or amenities
5. Offer fast deliveries
6. Customize products and services to customer needs
7. Profit in price competitive markets
8. Introduce new products quickly
9. Provide effective after-sales service
10. Offer a broad product line
11. Make rapid volume changes
12. Make rapid product mix changes
13. Make product easily available
14. Make rapid changes in design

They plan to stress the need to offer total quality: reliability and durability, consistency in low defect rates, production of high-performance products or **product amenities,** effective product support, good after-sales service, and dependable deliveries. Next comes the ability to customize products. Flexibility in design, volume, mix changes, and the ability to introduce new products are at the bottom of the list, along with the ability to compete in price competitive markets. Broadly speaking, the average European manufacturer's sequence of priorities is: quality, delivery dependability, price, and flexibility.

Table 6–5 shows how four industry groups—consumer packaged goods, basic industries, machinery, and electronics—view the priorities. There are striking similarities among them, but there are also important differences. Consumer packaged goods producers, operating usually in price competitive markets, emphasize delivery speed and price more; electronics, which is coping with rapid technological change and proliferation of new products, emphasizes customization and after-sales service to a greater extent; and the basic and machinery sectors emphasize product performance to a greater extent.

TABLE 6–5
Five Highest Competitive Priorities Per Industry

Consumer Packaged Goods (n = 26)	Basic Industries (n = 97)	Machinery (n = 39)	Electronics (n = 52)
Consistency	Consistency	Consistency	Reliability
Dependable deliveries	Dependable deliveries	Dependable deliveries	Consistency
Fast deliveries	Reliability	After-sales service	Dependable deliveries
Price	High performance	Reliability	Customization
Reliability	Fast delivery	High performance	After-sales service

Note: *n* equals the sample size.

What kind of objectives are these manufacturers setting for themselves? As one might have expected, improving quality, reducing costs, and increasing the productivity of the production process are the main themes in the top 10 objectives (Table 6–6.) However, three items on the top 10 list in Table 6–6 go beyond the traditional borders of manufacturing, which is significant and indicative of the increasing pressure to reduce and redefine the borders of manufacturing: Improvement of **interfunctional communication** (fifth most important objective) indicates the outbound reach of manufacturing towards these and other functions; improving vendor quality (eighth in the list) requires close interaction with vendors in addition to the definition of specifications and standards; and reducing new product development cycle time (ninth on the list) demands working more closely with R&D (research and development).

TABLE 6–6
Manufacturing's Objectives

Top 10

Improve conformance quality
Reduce unit costs
Reduce overhead costs
Increase delivery reliability
Improve interfunctional communication
Improve direct labor productivity
Increase throughput
Improve vendor quality
Reduce new product development cycle
Improve white-collar productivity

Bottom 5

Improve labor relations
Increase capacity
Increase range of products produced by existing facilities
Reduce number of vendors
Reduce capacity

TABLE 6–7
Intended Actions for the Next Two Years

Top 10

Linking manufacturing strategy to business strategy
Integrating information systems in manufacturing
Quality function deployment
Supervisor training
Worker training
Integrating information systems across functions
Management training
Interfunctional work teams
Developing new processes for new products
Statistical quality control

Bottom 5

Quality circles
Value analysis and product redesign
Hiring in new skills from outside
Robots
Closing, relocating plants

How are these objectives translated into action plans for the future? We asked our respondents to indicate the degree of emphasis they were placing on each of a list of 26 specific action programs for the period 1990–92 (see Table 6–7). Although it is not first on the list, training is clearly a key—training at all levels. The respondents estimate that 4 percent of their direct employees' time is spent on education and training (as compared to 65 percent on direct production tasks). This is about one day's training a month!

Five action plans in the top 10 reported in Table 6–7 have integration as their central focus. Closer links between manufacturing and the business strategy aim at enhancing the contribution of manufacturing for the competitive advantage of the company. Integration of information systems within manufacturing and across different business functions will require a horizontal integration with the rest of the company. Interfunctional work teams bring this integration to an operational level. Finally, the emphasis on quality function deployment (QFD) suggests further integration of the customer into production.

The intent to integrate is supported by other data from the survey: 32 percent of the respondents plan to pursue joint activities with both customers and vendors in the next two years; another 38 percent with customers, vendors, *and* other partners. On the whole, 80 percent of the respondents plan on some joint activity with their vendors. Only a small number of the respondents have no intention whatsoever of generating joint activities with others. These actions, which focus on the borders of manufacturing, are being emphasized more than many actions that relate to the core of manufacturing functions such as just-in-time, production planning and control, or quality control.

Those items at the bottom of the priority list in Table 6–7 also deserve attention. Recent research indicates that a better integration between design and manufacturing is essential for superior technology management. It is disconcerting to note, therefore, that **design for manufacture (DFM),** or **value analysis** and product redesign, score at the bottom of the list. The average European manufacturer does not consider integrated product design a priority for the immediate future. This may be a costly oversight, particularly considering the fact that the small group of manufacturers who did emphasize DFM obtained a good payback and placed it fourth on their list of priorities for the next two years.

What is the bottom line? We see a rather aggressive stance based on strengths in resources that have been built up in the recent years, and a restructuring of the production facilities. Competitive priorities continue to be stable and the manufacturing objectives that emphasize, in the first instance, quality and reliability, are consistent with them. The thrust of action plans has been aimed at improving the integration of manufacturing within the organization and generally removing the barriers between manufacturing and its partners inside and outside the company.

A somewhat simplistic model of the traditional position of manufacturing is presented in Figure 6–2. What this figure shows are the subtle ways in which manufacturing is normally insulated from its environment. Inventories insulate manufacturing from its suppliers and customers. Forecasting by sales departments, warehousing by distribution, procurement by purchasing, and product planning by marketing provide additional insulating layers. Laws and regulations can shield a factory from its environment. Meeting the labor laws or abiding by a union contract, for example, can give a factory a false sense of security, especially if management uses the legal or

FIGURE 6–2
Insulating The Factory From The Environment

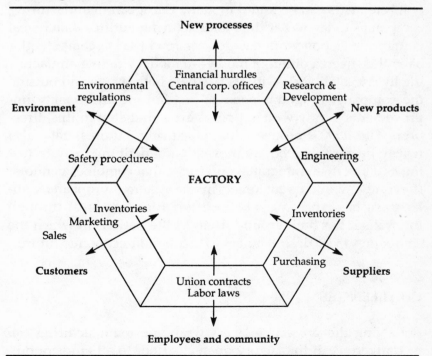

contractual obligations as a substitute for keeping track of changing labor conditions and safety and environment regulations. High hurdle rates for return on investments in new machinery can leave the existing machinery in the factory year after year; new products are the domain of engineering and can enter the production floor only when all details and specifications are clear; unions can form a buffer for direct interaction with employees; competitors are kept at a distance with trade secrets. Most of this is done to "protect" manufacturing, to make it easier to manage. The assumption has been that by providing more stability, manufacturing can be better fine-tuned and "optimized." The 1990 European Manufacturing Futures Survey shows that such an insulated manufacturing function is losing its relevance.

This extreme caricature of manufacturing as an isolated function has probably never existed in totality. However, some of its elements can be found in many European manufacturing companies today. What the Manufacturing Futures data reveal is that our respondents have understood that to compete globally they need a different model—a model where manufacturing lives in symbiosis with its partners both inside and outside the company. Programs such as just-in-time and quality improvements jointly with suppliers are a first step in this direction. The 1990 European Manufacturing Futures data also reveal the germ of a bolder move: that of defining manufacturing as a link in an integrated enterprise that combines vendors, the company, and its customers in one system. European manufacturing has to prepare to remove its borders—to break through the walls it has built around itself, as the nations in which the companies exist drop barriers to a common economic future.

CHALLENGES

Removing the protective layers wrapping manufacturing can be dangerous. If the factory opens its floor to all kinds of outside influences, the entire manufacturing system could become unstable and uncontrollable. To be successful in such a bold move, the foremost prerequisite is that a significant majority of company employees understand the direction. This is clearly not yet the case. Two thirds of the respondents to our survey indicated that only top and middle management understand the strategies, goals, and objectives of their company. This is insufficient. Removing the factory walls is a radical departure from past practices and needs cooperation and innovation by all.

Second, the interaction between manufacturing and product design must be substantially improved. As we indicated earlier, this interface is not being managed well in the European manufacturing companies. Other data in our survey suggest that we are still far from simultaneous engineering. Process engineering on the average starts when only half of the product engineering is finished. Integrated **CAD (computer-aided design)–CAM (computer-aided manufacturing)** is still far off for many of our respondents. Only half have both CAD and CAM and of these, only half have some kind of integration between them. Only 10 percent of manufacturing personnel have had any experience in design (and vice versa). Job rotation across functional borders is not yet common practice.

Third, the measures for manufacturing performance must be critically reexamined. Most of the traditional measures— output per hour, productivity, and utilization rate—discourage the removal of protective barriers. Other measures—on-time delivery, reduction of lead times (order, production, supplier, and other lead times), reduction of production batch sizes, reduction of inventories, speed of new product introduction— should be given higher priorities. All this calls for a new accounting system for manufacturing, which is a tall order. For example, as we have discussed, our data show that those European manufacturers who have tried to introduce activity-based costing (ABC) so far have not had a high pay-off (Table 6–2). The most likely explanation, in our opinion, is that the changes required are more difficult and time-consuming than expected. Therefore, much stamina is required to institute a new performance measurement system for manufacturing.

CONCLUSIONS

For a number of years, we have witnessed the race to improve the *core* of manufacturing operations. In early 1990, spurred by significant changes in the political and economic environment, manufacturers in Europe began to focus more on the *borders* of manufacturing—both internal borders with other functions and external borders with customers, suppliers, and others. Are they ready for the challenge of removing the walls around their factories? The 1990 European Manufacturing Survey data indicate that, even in this sample of high-performing manufacturers, much work still lies ahead. The hurdles are many and difficult to overcome. The process of exposing the shop floor to more and more outside influences is risky and is likely to meet much resistance in the company. The skeptics will be convinced only to the extent demonstrated by superior performance. A small slip is likely to retard the efforts considerably. So it is a journey that needs conviction and persistence as well as care and good ideas.

But the promised benefits are also great. An agile manufacturing function that is well integrated into the business and in direct contact with its outside partners will be a formidable competitive weapon. It will constitute a very defensible core competence that can change the nature of the competition in the industry.

Our data show that the advanced manufacturers in our sample are taking on this challenge. In a sense, this is the natural next phase for them. After revamping the factory, the efforts on improving the quality and reliability of delivery are leading them to the obstacles around the factory. Removing these obstacles will be a difficult and time-consuming task, but those who have concrete and careful plans to go after the obstacles are likely to show success soon.

There is a risk that the European manufacturers may not pay sufficient attention to this message. Things are going well, and the excitement of the 1992 single market, confounded with the opening of Eastern European markets, can discourage bold changes in manufacturing. With so much uncertainty and turbulence in the environment, it is quite tempting to increase the protective layers around production.

However, protective layers are costly in the long run. Therefore, now that things are going well, the company should risk going after developing new manufacturing capabilities. The challenge facing the senior manufacturing manager is to show that an open flow of goods, information, and people across the traditional boundaries of the manufacturing function is not only necessary to prepare for the future, but also profitable.

CHAPTER 7

JAPAN: CREATING CUSTOMER–DRIVEN FLEXIBILITY*

After Japan's emergence into world trade in the early 1900s and well into the 1980s, many methods and techniques for improving production systems were copied from manufacturers in the United States and Europe. However, these methods were so adapted to the Japanese management culture that they became distinctly Japanese, entirely different in practice from the same-name methods in their countries of origin. The Japanese, you might say, are expert benchmarkers. These transformations of American and European approaches to manufacturing production by the Japanese can best be exemplified by tracing over the last 50 years the implementation history of four management tools: industrial engineering (IE), quality control, just-in-time (JIT) production system, and materials requirements planning (MRP) system.

Industrial Engineering (IE)

Industrial engineering was the first of several methods and techniques to be brought to Japan. Industrial engineering in the United States was originally based on the study of workers'

---

*This chapter has been adapted from Professors Jinichiro Nakane and Seiji Kurosu, Waseda University, and Haruki Matsuura, Kanagawa University, "Creating Customer-Driven Flexibility: Executive Summary of the 1990 Japanese Manufacturing Futures Survey."

tasks and was aimed at the improvement of operations and processes following the early work of Frederick Taylor, H. L. Gantt, Frank and Lillian Gilbreth, and others. From the beginning, the discipline used a combination of work measurement and method improvement techniques. Work measurement was originally used to see where work could be improved. Later, especially in the United States, the practice of time standards was stressed as the most important reason for work measurement; methods improvement became secondary. In Japan, on the other hand, the introduction of industrial engineering was based fundamentally on work measurement for performance improvement. Further, in the United States, industrial engineering rapidly became the domain of the specialist. In Japan, in contrast, industrial engineering was not a matter only for specialists but had to do with all workers. Everyone was taught industrial engineering techniques by managers. Self-study approaches were also used in daily work as a way to practice these techniques. Small-group activities were developed to supplement and reinforce the importance of task mastery.

Quality Control (QC)

Following World War II, advisers from the United States, such as W. Edwards Deming and Joseph Juran, recommended that Japanese companies use statistical quality control techniques as a fundamental tool for improving productivity and quality. The receptivity demonstrated by the Japanese to this suggestion was far more intense than expected. Statistical quality control principles spread and the basic techniques were learned quickly. Other techniques like the plan-do-check-act (PDCA) cycle and fishbone tree were developed with the participation of various company's workers who were actively involved in the quality control movement.[1] In Japan, enthusiasm for qual-

---

[1] The PDCA cycle and fishbone tree are management tools suggested by Kaoru Ishikawa to accomplish total quality control. For more information, see Ishikawa and Lu, *What Is Total Quality Control?—The Japanese Way* (New York: Prentice Hall, 1985).

ity improvement programs led to the adoption of many other developments as well, including the creation of the profession of quality control, and the emphasis on self-inspection, self-evaluation, and other control systems. The predisposition of Japanese management to favor improvement activities that started with industrial engineering fostered the success of companywide quality control movements in Japan.

Just-In-Time (JIT) Production System

Toyota experimented with JIT production in the 50s based on its study of Henry Ford's production methods, and applied it widely in the 60s. But JIT thinking was not widespread in mainstream Japan until 1973. Only after the severe fuel shortages of 1973 gave new meaning to the concept of eliminating waste did this practice come into vogue. Then economic growth halted for a year, and it became apparent that it would not return to the pre–oil shock level soon. Competitive priorities shifted to holding or decreasing costs while producing relatively low volumes and increased varieties of products. JIT began to expand, beginning with original equipment manufacturers (OEM) companies; by the mid-70s, JIT programs had expanded into supplier companies. The expansion of JIT concepts continues today. Initially, the purpose of JIT was mostly to eliminate waste, but lead times were soon decreasing so drastically that the objective changed. Companies found that the flexibility they gained by shortening lead times and changeover cycles could be used as a strong competitive advantage in rapidly adjusting to customer needs. Lead times were cut not only in production but in all kinds of other activities, including product design cycles.

Materials Requirement Planning (MRP) System

Materials requirement planning (MRP) was developed in the United States for the integration of production systems. MRP organized informal procedures into a formal system and integrated the production management system. This provided better decision-making processes for production planning and control and was an important step toward the development of management systems to support high-level management. Our studies show that about 20 percent of Japanese manufacturing companies have introduced MRP systems through the end of the 1980s. MRP packages have been implemented in 500 to 800 companies, and the trend is growing. The introduction of MRP to Japan followed the pattern of industrial engineering and quality control. Specific departments were created for the management of the improvement of productivity and quality for both processes and organization. Until recently, however, the use of MRP systems in Japan has been limited. MRP systems were implemented independent of other production and control systems in use. The focus was on the information processing aspects of the MRP system rather than on its use as a control tool. Although MRP enables many key features of the Plan portion of the PDCA cycle, there is a trend in Japanese plants to apply it simply to the operational Check elements, for example, process management (including process improvement) and outside production delivery control. As a result, MRP has been used in Japan more as a decision support and analysis tool and less as a control system.

It is important to understand these transformations of imported production methods into tools that fit the Japanese approach to management before considering recent trends in Japan.

RECENT TRENDS IN JAPANESE MANUFACTURING STRATEGIES

In the 1990s, most Japanese companies are developing their business strategy with a primary objective of market share expansion by introducing new products into the market. The introduction of new products to the market not only has a positive impact on sales, but it makes employment security and the targeting of new markets possible. This market policy fits neatly with the emerging flexibility of Japanese manufacturing units, which are well able to support the production of high-quality, low-cost goods developed to meet individual customer needs. This convergence created the notion that customer-driven flexibility was possible. It enabled the Japanese manager to break loose from the basic assumption that bound the rest of the industrial world—that cost and flexibility are trade-offs.

In order to better understand Japanese manufacturing strategies, we now present the results of the Japanese Manufacturing Futures Project. First, we review where Japanese manufacturers have improved most in the recent past. Second, we describe the Japanese view of competitive priorities. Third, we look at the future plans of large successful Japanese manufacturers to improve their competitive capabilities in coming years.

PERFORMANCE IMPROVEMENTS THROUGH THE LATE 80s

Japanese manufacturers reported strong improvements in flexibility through the late 1980s. While the improvement on other competitive dimensions continued with steady speed, there was also a remarkable improvement in the ability to satisfy diversified customer requirements quickly. In other words, Japanese manufacturers moved with dispatch along the two critical dimensions of flexibility: variety and responsiveness.

Figure 7–1 describes the performance improvement of the average Japanese manufacturer in various areas during the period 1988–90. Significant improvements have been achieved in several dimensions of flexibility: set-up time reduction, number of products manufactured in a plant, delivery speed, and the speed of new product development. Comparative data from U.S. and European nations showed much slower rates of improvement in these areas.

FIGURE 7–1
Japanese Performance Improvement, 1988–1990

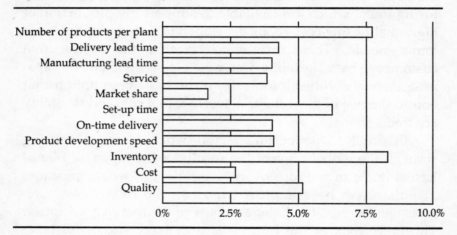

To account for this improvement in plant flexibility through 1990, we reviewed the action plans pursued by Japanese manufacturers in 1988. Table 7–1 shows **manufacturing lead time** reduction listed as the most important action plan for the following two years, followed closely by improving production control systems and procurement lead time reduction. Also highly ranked are improving design changes, new prod-

TABLE 7–1
Top 10 Action Plans in 1988

Ranking in Japan	Improvement Action Plans	Ranking in U.S.
1	Reduce manufacturing lead time	6
2	Improve production control system	10
3	Reduce procurement lead time	9
4	Improve time for design change	14
5	Increase ability to introduce new product	5
6	Giving workers broad range of tasks	17
7	Automation of operations	26
8	Computer-aided design (CAD)	20
9	Developing new processes for old products	22
10	Define manufacturing strategy	2

uct introduction capability, and computer-aided design (CAD). Coupled with worker flexibility, these action plans are intended to improve the flexibility of production systems. In contrast, U.S. manufacturers ranked as the most important improvement activities vendor quality, SPC, and worker safety, followed by manufacturing strategy and worker training. The relatively low rankings of the U.S. counterparts for those flexibility related activities stand in marked contrast to the Japanese priorities.

COMPETITIVE PRIORITIES FOR THE 1990s

Table 7–2 represents what Japanese manufacturers perceive as their competitive priorities for the next five years. Quality issues still dominate the list, but flexibility is becoming a more important dimension. Capabilities in design change and ability to introduce new products quickly are becoming more important to the Japanese manufacturers.

TABLE 7–2
Rank Order of Japanese Competitive Priorities in 1990

Order	Competitive Priorities
1	To provide reliable products
2	To make dependable delivery promises
3	To make design changes and introduce new products quickly
4	To offer consistent quality (low defects)
5	To make durable products and services
6	To provide fast deliveries
7	To offer low prices
8	To provide high-performance products
9	To provide after-sales service
10	To make rapid volume changes
11	To provide product supports
12	To offer wide range of products
13	To make product easily available (broad distribution)
14	To make changes in product mix quickly

FIGURE 7–2
Changes in Japanese Competitive Priorities, 1988–1990

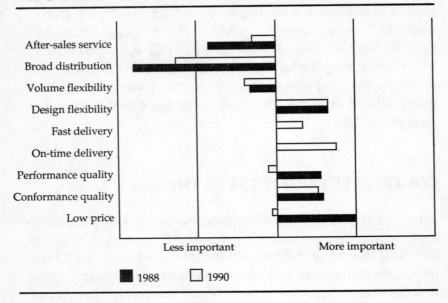

The increasing importance of flexibility in Japanese manufacturing strategy is seen more clearly when we compare the relative importance of each capability in 1988 and in 1990. Figure 7–2 shows that the Japanese companies see responsiveness, represented by fast delivery and on-time performance, as much more important in 1990 than in 1988. Fast response to changing market demand, along with design and product mix flexibility, will allow Japanese companies to compete more effectively in the future. In contrast, price and quality appear to be less emphasized than they were in 1988.

ACTION PLANS TO MEET CHANGING COMPETITIVE PRIORITIES

Competitive priorities should determine the manufacturing function's objectives. Table 7–3 reflects the key manufacturing performance targets as perceived by the Japanese manu-

TABLE 7–3
Top 10 Japanese Manufacturing Objectives in 1990

Order	Performance Targets
1	Reduce unit cost
2	Improve conformance quality (reduce defects)
3	Improve direct labor productivity
4	Reduce break-even points
5	Reduce manufacturing lead time
6	Increase throughput
7	Reduce new product development cycle
8	Increase delivery reliability
9	Increase delivery speed
10	Reduce set-up/changeover times

facturing companies. Clearly, cost and quality remain important performance priorities. However, it is noticeable that some measures closely related with flexibility are now high on the list.

Particularly interesting is the high rank given to the objective of reducing break-even points. This shows the intention of making factories smaller than they are. The business environment expected by Japanese managers includes lower demand volume for each product they offer; thus, in order to remain profitable, they will need to reduce break-even points. Also included in the high-ranking manufacturing objectives are various dimensions of time-related performance: manufacturing lead time, new product development cycle, delivery speed and reliability, and changeover time. Improved performance in these dimensions is critical for Japanese companies to enhance their responsiveness and flexibility.

The changing strategic direction and performance targets lead Japanese manufacturing companies to spend more resources on the action programs that are designed for flexibility enhancement. Table 7–4 summarizes action plans that will be receiving the most attention and resources by Japanese firms from 1990 to 1992.

TABLE 7–4
Top 10 Action Plans in 1990

Order	Action Plans
1	Integrating information systems in manufacturing
2	Developing new processes for new products
3	Investing in improved production inventory control systems
4	Developing new processes for old products
5	Integrating information systems across functions
6	Linking manufacturing strategy to business strategy
7	Computer-aided design (CAD)
8	Management training
9	Value analysis/product redesign
10	Quality function deployment

Developing new processes for old and new products is ranked high in Japanese action plans. Japanese manufacturers have been known historically for their continuing effort to improve their production processes. These efforts are necessary for the Japanese to respond to the increasingly diversified products they offer. Improving processes and making the best possible use of existing technology and facilities are critical first steps before considering an investment in new equipment.

Integrating information systems, both within manufacturing and across functions, is getting more attention from Japanese managers. In order to respond to fast changes in markets, it is imperative that marketing, engineering, and manufacturing functions communicate with each other effectively. Design engineers must understand what marketing and sales managers want in their new products, and manufacturing managers must configure production processes in order to deliver the products as defined by the design team. Information systems thus become the linchpin among various functions, and their integration is becoming more important.

Manufacturing strategy was also among the highly emphasized action plans. In fast-changing markets with a business strategy based on enhancing new product introduction and improving market share, Japanese managers perceive the importance of a strategic perspective to better manage their manufacturing function. More than just producing better quality at a low cost, the manufacturing function now has the task of being responsive and flexible. The tradition of continuous improvement is supplemented by more strategic leadership.

Specific techniques are called for to implement these strategic changes. Computer-aided design (CAD) is getting higher emphasis, along with quality function deployment (QFD) and value engineering. The purpose of CAD is to improve the efficiency in the design process. Efficiency in design is necessary to convert customer requirements into design and engineering specifications more accurately and promptly. The increased emphasis on this technique suggests that Japanese manufacturers are now looking more into a design flexibility that can be supported by their superior production flexibility capability. Value engineering is receiving renewed attention, and so is QFD, the technique that helps managers to better relate customer requirements and engineering specifications.

CUSTOMER–DRIVEN FLEXIBILITY:
EMERGING MANUFACTURING STRATEGY

In Japan, market demand for a wide variety of products requires an immediate response from manufacturers. Traditionally, Japanese manufacturing companies have pursued flexibility in order to respond to the market changes that lie ahead and they have achieved considerable success in flexibility in the production process. However, they appear to have reached a limit in their ability to improve responsiveness through manufacturing efforts alone. The results of the 1990 Manufacturing Futures Project show that slowly but surely, Japanese manufacturers are shifting their focus from shop floor flexibility to an interfunctional flexibility that is driven by customer requirements. Armed with superior production-driven flexibility, Japanese manufacturers are seeking customer-driven flexibility.

Customer-driven flexibility is a challenge that requires a significant amount of coordination between various functions, not just production capability. Leading manufacturers believe that the production system should be capable of handling multiproduct manufacturing requirements and that there should be more direct communication between manufacturing and customers. The wall between marketing and design engineering, and between engineering and manufacturing, is being broken down by these leaders as they strive for interfunctional agreement on the importance of customer-driven flexibility.

In response to this challenge, Japanese managers appear to be changing intentions and perspective in comparison with their traditional improvement efforts. First, their action plans have a more strategic and long-term orientation. More top-down innovative approaches are being emphasized than bottom-up improvement approaches. They appear to be looking for a few strategic leaps rather than relying only on step-by-step changes. Second, their improvement efforts are directed at the general management systems as well as improvement of hardware-based production systems. In the past, Japan has attached great importance to the shop floor and has succeeded in improving productivity and quality. More recently, we are seeing increased attention to upper level management and non-production functions. Clearly, Japanese manufacturers are building the future with a changed perspective, and it appears to be customer-driven flexibility.

THE LEADERSHIP CHALLENGE

Future challenges for Japanese manufacturers are more than just shifting demands. The global business environment is changing rapidly, and national competitiveness cannot be taken for granted. Technology development in manufacturing and non-manufacturing sectors is astonishing. Today's winner does not have a guaranteed position in the future at all. In order to maintain their current competitive position and meet future challenges and uncertainties, Japanese manufacturers need to consider the future environment much more than they have.

Several of the major changes in the manufacturing environment are foreseeable. According to results from our companion project, Manufacturing 2001, which was initiated to predict the business environment and potential manufacturing systems that might be able to respond to it, Japanese managers believe that the future will be much different from the present along a variety of dimensions. Some of the perceived changes in the manufacturing environment and success factors are summarized in Table 7–5.

Japanese manufacturers foresee lower economic growth rates, particularly influenced by maturing markets. They also see that more breakthrough changes are necessary in the future, not merely more incremental continuous improvement. Globalization is predicted as the key trend, and dealing with heterogeneous people and resources is perceived as

TABLE 7–5
Key Changes in Japanese Business Environments and in Competitive Success Factors

Past	Future
High economic growth	Low economic growth
Growth markets	Mature markets
Continuous development and adaptation of technology	More step-function (breakthrough) change in technology
Clear goal (catch up with the West); clear targeting	Indistinct goals
Domestic production; export marketing; domestic-oriented business activity	Globalization of all business activity
Homogeneous people with homogeneous ways of thinking	Heterogeneous people with many different approaches to both life and business
Pax Americana	Pax consortis (world order preserved by a consortium of advanced industrial nations)

an important task. The most significant challenge, however, appears to be in taking a leadership position, which is perceived by Japanese managers as much more difficult than simply playing a catch-up game. Japanese manufacturers have to overcome these upcoming challenges in order to maintain their global leadership. In this section, we present what were identified as challenging tasks in our Manufacturing 2001 project.

Innovations: More than Continuous Improvement

The era when mass production using inflexible hard automation could be competitive is nearly over in the advanced industrial countries. It may continue to be viable in newly industrializing countries, but manufacturers in advanced countries must provide more customization and a quicker response. Hence, customer-driven flexibility. Up to a certain point, flexibility can be attained by adapting existing equipment for faster set-ups, or by changing floor layouts to form production cells. Beyond that point, however, simple revisions of old-concept tooling and equipment can no longer compete against computerized equipment whose original design purpose is flexibility. New equipment will have automatic set-up capability built in. Equipment and facilities will be designed for fast layout change whenever needed. Computerized transmission of data and machining instructions will clearly surpass coordination by existing sight and signal systems.

There needs to be more than a continuous improvement philosophy to approach the new customer-driven flexibility requirements. Japanese manufacturers will have to change their traditional improvement-oriented management style to one that facilitates innovations throughout their business processes.

Managing the Hidden Factory

Already in many companies, the narrowly defined production function of material conversion consumes less than half the total resources required to serve the customer. Other processes, previously hidden from view and hence relegated to what J. Miller and T. Vollmann have called the "hidden factory,"* are now dominant. They include design, engineering, and logistics, which are becoming more critical in satisfying customer needs. This trend will continue, and the production function will be perceived of as a part of the total customer service process. The question arises, therefore, how this hidden factory should be managed. So far, most Japanese manufacturing managers have developed various procedures and technologies that were designed to improve shop floor activities. That is, until now, management systems have focused on production technology. There are indications now that companies will strengthen their management capabilities in marketing and finance to complement their manufacturing strength.

It will be a big challenge for Japanese managers to deal more effectively with the problems associated with non-manufacturing tasks. The proportion of direct costs will decrease, and the indirect costs will increase rapidly. Performance evaluation schemes must be developed to manage these hidden costs more effectively.

Management by Systems for Global Leadership

Japanese manufacturers have historically followed the technology leadership of Western competitors. That is no longer the case, as a number of Japanese industries have taken a leading position in industries such as automobiles and consumer electronics. Being a technical leader is a different experience than being a follower. The direction to travel and the goals to pursue are not as clear as when Japanese industry was simply trying to be a fast follower of the technology leaders. Furthermore, multinational operations seem to be here to stay. No advanced industrial country is a pocket of isolation. Scientific and technical advances may occur anywhere, and they may Ping–Pong

---

*"The Hidden Factory," *Harvard Business Review*, July–August 1985.

between several locations of the globe in the process of development. A major manufacturer needs a world presence in order to remain competitive. As always, aggressive leadership in the new direction by top management will make the difference in the fortunes of individual companies. They must overcome the corporate bureaucracies that impede the aggressive development of new ideas to meet new challenges.

The postwar era for Japan was supported by homogeneous resources, human as well as material, and it has allowed Japan to create a distinct system. In the future, however, the development of a management system has to deal with heterogeneity, especially in its human resource base. Until now, the Japanese management system has relied on the personal skills of the manager. This is not necessarily a bad approach, but when we consider the environmental changes ahead, a more systematic method of management has to be developed. For Japanese managers, computer-integrated manufacturing (CIM) and the total business integration system offer a paradigm on which they hope to build systematic approaches to non-production tasks.

FUTURE OF JAPANESE MANUFACTURING: FOUR AGENDAS

To cope with fast changes in global competitive environment, and to take advantage of Japanese management culture, managers participating in the project Manufacturing 2001 have identified some alternative future directions. We present four of them here.

Shift to R&D Leadership

Rather than being the leader in production technology and know-how, the emphasis will switch to new product innovation. Companies will be in the "research and development business," which will continuously generate new products for the market. This alternative allows Japan to avoid competing with companies from developing nations that could have better and lower cost resources for manufacturing. In order to shift completely to this type of strategy, Japanese companies will have to change their management system to emphasize innovation and creativity rather than continuous improvement.

Become "New Style" Low-Cost Producers

Japanese manufacturers can use their rich experience in continuous improvement and develop further automation by simplifying and integrating systems. When this cannot be done, they can opt to move production offshore and exploit low-cost resources. Different from the traditional definition of low-cost producer, which is based on low production factor costs such as labor and land, this new approach will emphasize sophisticated process control and a high degree of automation. This alternative is certainly possible considering the past achievements of Japanese industries, but could be perceived as less desirable because of the high degree of competition and the low value-added nature of this type of business.

Take Advantage of Flexibility

Flexibility in a narrower sense is the rapid and efficient handling of product mix and volume changes in production. In a wider sense, flexibility represents the capability of quickly deploying both new products and new technology. Japanese know-how in flexibility, coming from JIT, TQC, and total factory automation, can be a formidable competitive weapon. Challenges for this alternative would come from the need to greatly reduce lead times for both design engineering and manufacturing engineering and still operate efficiently. In order to realize this capability, innovative actions, in addition to the traditional step-by-step improvement (*kaizen*) programs, need to be added in Japanese management culture. This strategy can be combined with a shift to R&D leadership easily.

Develop a Multinational Management System

Japanese companies manage multinationally now, but they do not consider themselves to be doing it very well. Great improvement on this front is a tough challenge. Japanese firms need to be more persistent and accumulate much more trial-and-error experience in this challenging task. Rather than giving up after short-term trials and failures, they should keep their foreign operations initiatives and experiment to find long-run success. Japanese managers have been successful at home with their informal management methods based on personality. To succeed in global operations with largely heterogeneous resources, they need to develop more formal management methods. Therefore, the development of the capability to manage by formal systems is seen as a big challenge for the multinational business alternative.

Any individual Japanese company may pursue any one of these four scenarios, or one not listed. Most Japanese managers tend to favor the flexibility alternative—smaller, more flexible factories with additional reduction of lead times for domestic Japanese manufacturing. Regardless of which alternative they choose, integrative computer systems are seen as one of the keys to the manufacturing vision of the future. As discussed earlier, Japanese managers tend to operate more by informal methods than by formal ones, creating harmony from chaos. However, formal computerized systems depend on clear definitions of the roles and responsibilities of different groups and different individuals, which have not been strong points in most Japanese organizations. Overcoming this management culture to enable the use of information systems could be the most challenging task assigned to Japanese managers looking toward the 21st century.

CHAPTER 8

MANUFACTURING AROUND
THE PACIFIC RIM*

The Pacific has been the focus of manufacturing activity throughout the 1970s and 80s. Much of the attention has gone to Japan, but Korea, Taiwan, and other Asian "tigers" have also played an important role. In this chapter, we examine five countries that represent the broad spectrum of players on the Pacific Rim—Australia, Korea, Mexico, Singapore, and New Zealand.

The choice of these five countries is a reflection of several factors. Geographically, they encircle the Pacific Ocean, and thus have an important physical relationship to the two industrial powers, Japan and the United States, that dominate Pacific trade. The second factor is that they are representative of the many other important nations from this region. Indeed, these countries reflect the diversity of manufacturing to be found in nations around the world. The contrast between a vigorously aggressive Korea in full transformation, an ambitious Mexico, a successfully established Singapore, and an Australia and New Zealand struggling for industrial survival is vivid. It

*In addition to being the authors or coauthors of the reports adapted in this chapter, we wish to acknowledge Kee Young Kim of Yonsei University in Korea, Norma J. Harrison of the University of Technology in Sydney, and Lawrence Corbett of Victoria University in New Zealand for their help in putting together the conclusion on the Pacific Rim countries for this chapter.

References for individual sections of this chapter are cited with the appropriate sections.

contributes to an understanding of the range of manufacturing strategies that exist and the national circumstances that contribute to their existence.

Australia is a physically large country with a relatively small population and domestic market. For the last 30 years, manufacturing has been ignored in favor of primary production, mining, and the service industries. The paucity of value-added production and increasing protection in traditional export markets has led to a balance-of-payments problem over the last decade. The Australian government does not have an accepted overall strategy to address the problem. The response has been left largely to those manufacturers who see a long-term future in manufacturing. Unfortunately, they are saddled with rising input costs resulting from a large government sector.

New Zealand is in a worse position than Australia because it does not have the vast array of natural resources bestowed on its western neighbor. It is able to produce agricultural commodities very cheaply and in large quantity, but increasing protectionism in its traditional export markets has resulted in a severe balance-of-payments problem here, too. Large areas of the economy are subsidized, but recent governments have been stripping away protection in many sectors. The manufacturing sector is struggling because the domestic market is depressed and probably will be for some time to come.

South Korea, which was born as a country after the war with its communist northern neighbor, had been the agricultural part of the former combined country, while the north had been the industrial part. Hence, Korea had to build its manufacturing sector through direct government support and intervention. In this respect, it could be compared to Taiwan. The Korean government had to borrow heavily to enable industries to become established. These debts have been largely repaid. It is expected that the government will continue to provide direction on the steps manufacturers should take to remain competitive.

Singapore is a fairly small island with limited resources apart from its people. The manufacturing options available to it are aided by the ease of transport from a very efficient and widely used seaport and airport. The government, since political independence from Britain, has been very determined in the direction it gives the economy and this will likely continue in those industries that are seen as appropriate.

Mexico is a moderate-sized country with considerable natural resources and a relatively large, but poor, population. The country has suffered in the past from unstable government and rampant inflation but that now seems to be largely in the past. There is a much clearer focus on the future. The free trade pact that the United States and Canada propose with Mexico will provide a large market that was not available earlier. Manufacturers can concentrate in further utilizing the large pool of relatively low-cost labor.

The information on each country described in this chapter is presented as a series of key facts revealed by analysis of the 1990 Manufacturing Futures Survey followed by a brief synopsis for each country. More detailed reports on these countries can be obtained by contacting the authors from each country, or in the United States, from the Boston University Manufacturing Roundtable.

Tables 8–1 through 8–4 summarize the performance, competitive priorities, future action plans, and experience with action plans for each of the five Pacific Rim countries discussed in this chapter. These tables have been composed so that the manufacturing strategies in these countries can be generally compared with those from Europe, Japan, and the United States detailed elsewhere in this section and in Part 3 of the book.

AUSTRALIA: RESTRUCTURING FOR COMPETITION[1]

The 1980s were a period of hope for Australian manufacturing. Key terms such as *resurgence, world competitive manufacturing,* and *competitive advantage* were introduced into the vocabulary by

---

[1]This section has been adapted from Professor Norma J. Harrison, University of Technology, Sydney, "Restructuring for Competition: Australian Manufacturing Strategies in the 1990s," with her permission.

TABLE 8–1
Annual Rates of Improvement for the Five Pacific Rim Countries (In Order of the Measures with the Highest Annual Rate)

Rank	Australia	Korea	Mexico	New Zealand	Singapore
1	Overall quality 15.00%	Variety of products 13.70%	On-time delivery 7.70%	Profitability 8.80%	Customer service 17.00%
2	Customer service 13.00%	Speed of new product development 11.30%	Equipment changeover/ set-up time 6.70%	Inventory turnover 7.30%	Market share 16.00%
3	Inventory turnover 12.00%	Set-up time 10.90%	Overall quality as perceived by customers 6.30%	On-time delivery 5.70%	Overall quality 14.00%
4	On-time delivery 11.00%	Overall quality 10.50%	Market share 5.70%	Customer service 5.30%	Product variety 14.00%
5	Set-up time 10.00%	On-time delivery 9.40%	Customer service 5.00%	Variety of products producible 5.30%	Inventory turnover 12.00%
6	Market share 8.00%	Customer service 9.00%	Inventory turnover 4.30%	Overall quality 5.00%	Speed of new product development 10.00%
7	Product variety 7.00%	Profitability 8.70%	Variety of products producible by manufacturing 4.00%	Speed new products 4.30%	On-time delivery 9.00%
8	New product speed 5.00%	Manufacturing lead time 7.10%		Changeover time 4.30%	Set-up time 7.00%

TABLE 8–2
Competitive Priorities, 1990

Rank	Australia	Korea	Mexico	New Zealand	Singapore
1	Conformance quality	Conformance quality	Price	Conformance quality	Quality conformance
2	On-time delivery	On-time delivery	Quality (low defects)	On-time delivery	Reliability
3	Reliability	Delivery speed	Delivery speed	Reliability	On-time delivery
4	Price	Reliability	On-time delivery	Price	Delivery speed
5	Delivery speed	New products	Service—ability to customize	Delivery speed	Quality (performance)

TABLE 8–3
Emphasis on Future Action Programs to Be Implemented in 1990–1991

Rank	Australia	Korea	Mexico	New Zealand	Singapore
			Five Most Emphasized		
1	Worker training	Quality function deployment	Link manufacturing strategy to business strategy	Link manufacturing strategy to business strategy	Link manufacturing strategy to business strategy
2	Link manufacturing strategy to business strategy	Supervisor training	Worker, supervisor training	Supervisor training	Worker, supervisor, and management training
3	Supervisor training	Management training	Statistical quality control (SQC)	Quality function deployment	Integrate information systems in manufacturing
4	Multiskilling/award restructuring agreements	Reconditioning physical plants	Management training	Integrate information systems in manufacturing	Quality function deployment
5	Giving workers a broader range of tasks and responsibilities	Worker training	Giving workers a broader range of tasks and responsibilities	Management training	Integrate information systems across functions

TABLE 8–3 (continued)

Rank	Australia	Korea	Mexico	New Zealand	Singapore
			Five Least Emphasized		
1	Robots	Closing/relocating physical plants	Robots	Hiring new skills from outside	Closing/relocating plants overseas
2	Closing/relocating physical plants	Hiring new skills from outside	Closing/relocating physical plants	Computer-aided manufacturing (CAM)	Robots
3	Computer-aided design (CAD)	Robots	Design for manufacture	Computer-aided design (CAD)	Hiring new skills from outside
4	Equal opportunity/ affirmative action for women	Manufacturing reorganization	Computer-aided design (CAD)	Closing/relocating plants	Flexible manufacturing systems
5	Hiring new skills from outside	Computer-aided manufacturing (CAM)	Computer-aided manufacturing (CAM)	Robots	Computer-aided design (CAD)

(concluded)

TABLE 8–4
Effectiveness of Past Action Programs Introduced in 1988–1989

Rank	Australia	Korea	Mexico	New Zealand	Singapore
			Five Most Effective		
1	Manufacturing reorganization	Reconditioning physical plants	Closing/relocating physical plants	Link manufacturing strategy to business strategy	Supervisor training
2	Develop new processes for old products	Statistical quality control (SQC)	Reconditioning physical plants—JIT	Manufacturing reorganization	Worker training
3	Link manufacturing strategy to business strategy	Worker training	Supervisor training	Improve inventory control systems	Quality function deployment
4	Management training	Supervisor training	Computer-aided manufacturing (CAM)	Quality function deployment	Link manufacturing strategy to business strategy
5	Reconditioning physical plants	Management training	Computer-aided design (CAD)	Giving workers more tasks/responsibility	Management training

TABLE 8-4 (continued)

Rank	Australia	Korea	Mexico	New Zealand	Singapore
			Five Least Effective		
1	Equal opportunity/affirmative action for women	Activity-based costing	Robots	Closing/relocating physical plants	Closing/relocating physical plants
2	Robots	Robots	Giving workers more tasks/responsibility	JIT	Robots
3	Link structural efficiency plans and strategic management	Computer-integrated manufacturing systems (CIM)	Statistical quality control (SQC)	Quality circles	Hiring new skills from outside
4	Activity-based costing	Computer-aided manufacturing (CAM)	Design for manufacture	Design for manufacturability	Value analysis/product design
5	Multiskilling/award restructuring agreements	Hiring new skills from outside	Activity-based costing	Robots	Integrating information systems across functions

(concluded)

business schools, policymakers, and management consultants. While these concepts are useful and should not be forgotten, the 1990s will be a time for testing conceptual frameworks for their effectiveness in dealing not only with industry survival, but also with setting structures within which growth is possible. The key words of today in Australia are *rationalize, consolidate*, and *manufacturing restructuring*.

Strategies

Business strategies in Australian manufacturing organizations have taken a marked redirection. In the 1980s, manufacturers placed a higher priority on maintaining market share in existing markets. In the next decade, the business strategy of nearly two thirds of the respondents is expected to move toward building market share, with a third struggling to survive. Recognizing that a limited domestic market may necessitate increased exports for a viable operation, manufacturers planning increased exports are looking away from their traditional European and American markets and toward Japan and Southeast Asian niche markets.

Exports and offshore production were at dismally low levels in 1990, with about 85–90 percent of sales and production concentrated within Australia. This pattern of offshore activity conflicts with the intention to expand in the Pacific. More seriously, Australia's dependence on offshore purchases does not seem to be declining. The proportion of offshore to total purchases is about 24 percent for those in the survey. The net effect of these trends will do little to alleviate the already troubled balance of payments issue in Australia in the 1990s.

Competitive Priorities

Australian manufacturers see themselves competing primarily on quality and delivery standards through 1995. Price, and therefore cost levels, have also become an important strategic priority. There is, however, a wide "competence gap" between these perceived competitive priorities and the business units' current ability to achieve them.

A number of barriers hinder reduction of the gaps. The shortage of skilled workers is a major impediment to productivity increases and quality manufacturing. In addition, prominent internal barriers include restrictive work practices, the persistence of old cultures and lack of shared vision, increasing labor and overhead costs, the reluctance to invest in technology, slow product development, and the lack of integrated information systems and computer systems for design and manufacturing. Perceived external barriers include increasing material and energy costs, government policy and its effects on the economic climate, increasing global pressures, and a small domestic market.

Manufacturers realize that if they ignore their workplace problems, they risk losing market share or, worse, bankruptcy. There are conflicting views about how change should be effected. However, there is a shared acceptance that restructuring and efficiency are goals that must be pursued if Australia is to survive as an industrialized country. Employees at all levels should be involved in structural efficiency negotiations for any success in implementation. This is not the case at present. The key element in the restructuring of jobs is the training necessary to enable individuals to perform more effectively. It is uncertain whether manufacturers are presently satisfied with the level and quality of training provided.

Performance Improvement

Over the 1987–89 period, a wide range of performance improvements was recorded in the different industrial sectors. Increases of 8 percent in market share and 31 percent in profitability were reported by the major manufacturers surveyed. Although improvements were also registered in quality standards, customer service, inventory turnover, reliable delivery, and set-up time, the rates of improvement were not as high as those recorded in the 1986–88 period. In fact, lead times registered no improvements.

Reductions in the unit cost of production averaged only 2 percent in the 1987–89 period. In some cases, this reflected the initial costs that accompany the implementation of new technology. This is consistent with some manufacturers' claims that their most significant accomplishments during the 1980s were the ability to introduce new technology, capital expansion, and retooling in a worsening economic environment, as well as the successful introduction of quality control systems.

Past Action Programs

Action programs and plans that were of high priority between 1986 and 1989 concentrated mainly on improving existing plant systems rather than technological enhancements in plant and equipment, research and development, and developing design capability. These plans included the improvement of skills and management-labor relations, the enhancement of production and inventory control systems, and quality improvement.

In the 1990 survey, we note a change in high-pay-off action programs, with structural reorganization, the development of new processes, plant upgrades, and ensuring that manufacturing strategies are compatible with corporate strategies leading the list. Programs that have had trouble being implemented with high pay-offs include those that require relatively more capital investment, longer implementation times, and detailed management-labor negotiations. The message from the survey respondents is that success in manufacturing does not come easily and seems to require investment in a wide portfolio of programs, rather than concentrations in individual programs.

Future Action Programs

The strategies and action plans for 1990–91 were examined to see whether they addressed the issues of the 1990s and the objectives set in the manufacturing plans. The objectives to improve conformance quality, reduce all costs, and improve employee attitudes and morale at all levels of the organization are reflected in the major plans.

The need for effective manufacturing strategies has at last been realized, with emphasis on linking manufacturing strategy to business strategy. In formulating the action plans for 1990–91, manufacturers seem to have taken into account the lessons of the 1980s. Exceptions to this arise in two very important areas—product design and investment in new technology—that are recognized as crucial activities but take a much lower priority in the present action plans.

Conclusion

It is therefore clear that to succeed in today's environment, Australian manufacturing has to radically restructure. This restructuring should encompass process technology, manufacturing facilities, manufacturing planning and control systems, and organizational structures. Once manufacturing strategies have been specified, restructuring policies and decisions can be made that are consistent with them. It is crucial that these parts be compatible with each other to make the restructuring process successful.

KOREA: FROM LOW COST TO HIGH FLEXIBILITY[2]

Korea's past success has been built around a conventional corporate strategy of exporting and original equipment manufacturing (OEM). As such, corporate strategies place a premium on high-volume and low-cost manufacturing. The competitive priorities were simple: manufacture to conform to quality specifications as set forth by the OEM buyer, and offer low prices and dependable delivery. This once simple and secure strategy is rapidly disintegrating as OEM customers become more

---

[2]This section has been adapted from Professors Kee Young Kim and Dae Ryun Chang of Yonsei University, "From Low Cost to High Flexibility: Manufacturing in Transition," with permission.

demanding and as Korean companies move upstream into OEM markets. Korean companies now face a new, more rigorous business environment that renders conventional manufacturing strategies inadequate.

Korea's rapprochement with socialist and former socialist countries presents new challenges and opportunities for Korean businesses. At the same time, trade relations with the United States and European countries are less harmonious than in the past. Moreover, regional economic integration threatens access to Western markets. Along with the newly found democratization in Korea has come social and labor unrest with rampant wage hikes and lower productivity, lower quality, and lower dependability of delivery.

These overall changes in the business environment call for structural adjustments. The key to restructuring Korean manufacturing will be to develop high flexibility–related capabilities. Korean manufacturers need flexibility to develop and market new products to promote growth, flexibility to penetrate new markets and thus lessen dependency on overcompetitive existing markets, and flexibility to accommodate changes in manufacturing without an exorbitant increase in costs.

Strategies

To respond to the new business reality, Korean companies have become more aggressive in building market share, even in comparison to their North American, Japanese, and European counterparts. This is true across all the industries surveyed. More and more, Korean companies are going offshore not only to market their products but also for their production and purchasing. Korean companies must open new markets, and overcome high labor costs at home as well as the increase of trade barriers, all of which necessitate direct foreign investment in production facilities. Conversely, the high dependency on foreign raw materials weakens Korean companies. Backward integration and localization of key materials is therefore a significant strategy for the future.

Competitive Priorities

The most important competitive capabilities as perceived by Korean manufacturers include conformance quality, on-time delivery, and speed of delivery. The position of quality is not surprising given the attention paid to quality as a strategic competitive weapon in recent years in Korea and elsewhere. One notable change is the drop in the importance of low price from 3rd in 1988 to 20th in 1990. Of course, low price is not as attainable as in the past because of increases in labor costs. Moreover, the switch from high volume–low cost to low volume–high variety diminishes the importance of price as a key competitive capability. Third on the list of important competitive capabilities in 1990 is delivery speed, which indicates the difficulty Korean companies have encountered in recent years meeting schedules because of labor work stoppages.

There is a large gap between the perceived importance and the actual level of competitive capability. These differences are much greater than we find in U.S., Japanese, and European survey responses. These differences can be partly explained by the pervasive perception that the competitiveness of Korean manufacturers has been dramatically eroded because of the fundamental political and economic shifts mentioned earlier. The gaps can also be explained because of the implementation lag associated with the shift to new manufacturing strategies. The competitive strengths are seen in conformance quality, on-time delivery, and reliable quality, while weaknesses are seen in low price and flexibility (design changes and product mix).

Performance Improvement

Korea's manufacturing performance shows not only dramatic improvement over the last three years, but also greater improvement than that shown in the United States, Japan, and Europe. The greatest improvement has come in the factors related to flexibility; that is, high variety production capability, speed of new product development, and set-up time reduction. These findings suggest that Korean companies are converting from conventional low price OEM-based high-volume manufacturing systems to those that can accommodate market demand for higher quality and variety at a rapid pace.

Past Action Programs

Korean manufacturing companies have been retooling their processes through rationalization. Work method and work environment improvement, statistical quality control, new processes for existing products, and new cost-estimation systems are examples of the tools to improve, and lead the list of high-pay-off actions. The underlying rationale is to set up the basis for developing and manufacturing new products. There are interfunctional linkages being established between manufacturing and other departments, and also vertical linkages between manufacturing and corporate strategies. Because these new strategies are difficult to implement without the cooperation and understanding of key personnel such as workers, plant supervisors, and managers, the education and training of these corporate personnel is critical. On the list of least effective action programs we find robotics, CAM, and CIM. While these programs may be important, the level of Korean technology in these areas is as yet too premature for them to show positive results.

Future Action Programs

Korean manufacturers picked as their most important future action improving the ability to reflect customer needs in products. This signals a stronger marketing orientation in Korean manufacturers. Combined with a high emphasis on developing new processes for new product development, we see a consistent picture of Korean manufacturers moving gradually away from their old high-volume, low-price strategy. We can also observe that reconditioning physical plants, supervisor training, management training, worker training, SQC, JIT, and value engineering are actions aimed at fulfilling the scientific improvement of manufacturing and also overcoming many of the current manufacturing problems. Automation through robotics, reorganization of manufacturing organization, and CAM should, of course, be important objectives but their low ranking indicates the perception of Korean manufacturers that their successful implementation will not happen overnight.

The integration of manufacturing information systems with those of other functions ranked only 10th in importance. As compared with the much higher position of this plan in the United States, Japan, and Europe, it shows the lack of awareness of the need for an integrated management system. Product development and process technology improvement are still at a low level of emphasis.

Korean manufacturing must adjust to the new environment of competition. Progress has been slow and includes some key problem areas. The manufacturing cost-to-sales ratio for Korean manufacturers for the 1990 survey is 80.7 percent (84.4 percent in 1988) and is expected to decrease to 77.7 percent in the next two years. If we examine the cost structure of Korean companies in 1990, materials cost is 66 percent, direct labor is 10.8 percent, energy 3.6 percent, indirect manufacturing costs 19.4 percent, and as compared to the results from 1988, materials costs are decreasing while those for labor and overhead have increased. Korea still carries about a 10 percent higher materials burden than other countries, but labor costs are still relatively low. Nevertheless, because of past reliance on building competitiveness on "low cost through low wage rates," the labor disruptions of the previous two to three years have had a negative impact. As a result, many manufacturers are pursuing labor substitution through factory automation. Even so, the large materials burden has contributed to cost increases for Korean manufacturers because key parts and components are imported from abroad, and second, small to medium-sized suppliers are labor intensive and inefficient. To reduce the materials costs, localization will be critical as well as developing R&D capability in materials development.

Organized labor activity is perceived as the most critical among the key internal corporate obstacles. It has been the most difficult manufacturing obstacle since the democratization of Korea's political economy in the last two to three years. We should note, however, that many companies that have attended to labor concerns have had a measure of success. As compared to the United States and Japan, Korean companies are highly centralized, with most of the key strategic and business planning conducted by top executives and middle managers and only limited participation and input by supervisors and workers. This failure to motivate the workforce into sharing and understanding the corporate vision, strategies, and business plans may be a factor in labor's unrest.

Technology development is one of Korea's key weaknesses. Across most of the industries surveyed, manufacturers rated themselves quite inferior in terms of needed technology. This weakness is particularly evident in new components and materials, while it is relatively better in manufacturing processes and automation technology. The former result can be explained by the high dependency on foreign sources for these products and their high cost. The spending on technology, as measured by R&D to sales, is low compared to manufacturers in the United States, Japan, and Europe. Technology development requires integration and teamwork between manufacturing, marketing, engineering, and other key functions, as well as a close information network that is sorely absent in Korean manufacturers and is not even mentioned among the top priorities in their action programs.

Conclusion

The results of the 1990 Korean Manufacturing Futures Survey show that Korean manufacturers are undertaking fundamental shifts in their corporate strategies, manufacturing priorities, and future action plans. Many companies are attempting to change their traditional low-cost, high-volume manufacturing strategies to those that involve higher quality, delivery, and flexibility. Nevertheless, the road to this higher ground of value-added production is not smooth.

Korean manufacturers currently lack the ability to realize their new competitive priorities. Moreover, a more systemwide orientation is lacking to accommodate flexibility-based strategies. Ultimately, the key to Korean companies will be their ability to instill a new manufacturing philosophy that balances the pursuit of low cost and high flexibility. In essence, Korean companies must lessen the trade-offs involved in attaining low cost and high flexibility and instead make these two goals mutually compatible and complementary strengths.

MEXICO: TOWARD A NORTH AMERICAN COMMON MARKET[3]

Until the early 1980s, the industrial development of Mexico was centered on internal market growth and the substitution of imported products with local ones. This policy generated little competition, few suppliers, highly integrated corporate structures, low volume, and little international competition. Then, in 1985, the Mexican government decided to open the economy and join GATT (General Agreement on Tariffs and Trade). This resulted in a rapid change in the industrial structure of the country. Following its decision to join the international community, the Mexican government proposed the creation of a common market between Canada, Mexico, and the United States. The resulting negotiations between the three governments have already prompted numerous joint ventures, acquisitions, and alliances within industry.

The sudden upsurge of interest in the free trade agreement is related to the worldwide movement towards global trading and operations—a move in which national boundaries are becoming less important for the design and operation of transnational firms. The economic revolution created by the

---

[3]This section has been adapted from Professors Rafael Arana, Miguel Leon, and Carlos Cosio of IPADE Mexico, "Manufacturing Futures Project: Manufacturing Strategies," with permission.

proposed common market will affect Mexico in two ways. Mexico can competitively produce to assist the United States and Canada in increasing export volume, while simultaneously benefiting from their sophisticated technology and capital markets. The most affected industrial markets in Mexico will be those that are capital intensive. These sectors will undergo continuous transformation.

For years, Mexico operated as a traditional developing country. Many state-owned companies were established to ensure output and employment without the worry of direct competition. These policies have largely been reversed and the Mexican government has chosen to privatize telephone companies, airlines, and banks. Products designed abroad but requiring a significant amount of low-cost labor will continue to suit Mexico's economy and abilities. However, Mexico is also working to become a significant player in certain sectors of technologically based competition.

Strategies

Mexican companies are preparing strategies to face implementation of the free trade agreement with Canada, Mexico, and the United States. The free trade agreement is forcing Mexican companies to build market share as a stronghold against foreign competition and as a negotiating asset for the establishment of joint ventures. When Mexico's economy was closed to international investment, market share was a function of installed capacity. Now, however, market share is based on quality, delivery, and cost.

Exports in Mexico are on the rise, a very important trend for Mexican companies. Production outside Mexico is growing at a lower pace; but it is moving and will be an area of increasing interest for Mexican firms. Outsourcing is also expected to increase substantially. Between 1989 and 1992, foreign sales are expected to rise from 5 percent to 14 percent, while foreign purchasing is expected to rise from 21 percent to 32 percent of total purchases.

Sixty-five percent of purchased products consist of proprietary products purchased externally. This illustrates the interdependency among producers and vendors. The interdependency is forcing more partnerships among producers and vendors. It is important to emphasize that 16.8 percent of purchased components were designed specifically for business units. This figure is expected to rise in the near future as competition increases.

Companies surveyed indicated that top and middle management have a good understanding of overall plans, strategies, and goals within the business unit. Supervisors and workers, however, appear to have very little understanding of strategic issues. Presently, everyone is involved in the quality process. Corporate mission and value statements are being published and distributed throughout the organization. This is producing positive results for labor-management relations. Due to increasing competition, companies are very concerned with communicating these goals, plans, and strategies to everyone in the company to avoid a lack of collective focus in the competitive race.

Competitive Priorities

The main concerns facing Mexican businesses are increasingly smaller profit margins, faster changes in process technologies, capital scarcity, and rapid product changes. Managers are finding it necessary to rationalize—production lines with fewer products per line. They are exporting the products in which they are highly competitive but importing lower volume products. Forecasting future product types and demand volume has become increasingly difficult. Managers are also having difficulty converting their production process technologies to the new requirements for their changing product lines.

As a consequence, Mexican industry is focusing on four important competitive abilities to maintain a competitive edge: price, quality, delivery, and service. Most of the companies surveyed believe that price is their number one competitive strength, with significant potential for improvement. Along with price, quality remains a top priority for Mexican firms in spite of the improvement that has occurred in this area over recent years. With customers pulling, clients in the production-distribution chain are pressing suppliers for more frequent and on-time deliveries. This is paying off as efforts are being made by every company to improve delivery service. Since interest rates are still very high in Mexico (30 percent annually), frequent and on-time delivery helps to free working capital. Major effort is now being focused on fulfilling customer needs. This is an area that all companies report looking into very aggressively as a way to build loyalty and increase the number of permanent customers.

Performance Improvements

The results of the survey indicate that delivery and quality show major improvements, with most of the companies heavily committed to quality programs and customer focus. Profitability did not show improvement because of the capital investments necessary for plant upgrades and due to the intense competition, both local and foreign, that is forcing companies to build market share. Process flexibility also ranked high in the area of improvement as companies devoted many resources to speed up response time, reduce inventories, and lower working capital requirements. Customer service increased 11 percent as a result of process quality, competition, and the conviction of top management that an absolute devotion to fulfilling customer needs is the only way to survive.

Past Action Plans

The main objective for Mexican manufacturing is the reduction of overhead costs. Companies are adopting a leaner organizational structure as corporate staff is drastically reduced. To control material costs, the surveyed companies are relying on reduced scrap and increased yields. This area provides a golden opportunity for cost reduction.

The major pay-offs over the past two years have come from action plans such as closing and relocating plants. Focused plants are replacing the plants with broad product lines. Likewise, the reconditioning of plants has provided competitive advantage. Manufacturing no longer appears to be a neglected area. Value analysis, quality function deployment, and interfunctional teams have worked very well. On the other hand, surprisingly enough, SQC (statistical quality control) has proved ineffective because of inadequate worker training, lack of supervisor understanding and support, and overly high corporate expectations.

Future Action Programs

To achieve a higher productivity, companies are looking towards their workers. Morale and manufacturing culture has also been rated as top priorities for competitive weapons; this indicates a new focus dimension for manufacturing in the immediate future. The maintenance of quality improvements and lower costs will be a core objective for survival, and interdepartmental communication is urgently needed in order to achieve a better market response. Manufacturing direct labor currently spends 65 percent of its time on direct production tasks. Solving and analyzing problems is a key area for improvement and is slowly but surely becoming a standard task for direct labor on Mexican factory floors.

NEW ZEALAND: BATTLING FOR SURVIVAL[4]

New Zealand manufacturers are in the midst of a struggle to remain competitive against domestic and foreign competition, to control costs, and to meet rising customer expectations of quality. The typical manufacturer has been vigorously shaken

---

[4]This section has been adapted from Lawrence M. Corbett of Victoria University "Manufacturing Strategies: Executive Summary of the 1990 New Zealand Manufacturing Futures Survey," with his permission.

by recent market conditions and regulation changes. These have forced them to undertake significant reorganization and upgrading of the manufacturing function and many see this as their most significant accomplishment of the 1980s.

Strategies

Despite their weaknesses, particularly those associated with competing on price, New Zealand manufacturers believe they need to expand to prosper in the future. Sixty-nine percent of the respondents plan to build market share over the next two years. Some of this growth will come from increased offshore sales and from moving more production offshore.

Overall, New Zealand manufacturers who responded to the survey seem confident about the next two years. They intend to follow a coherent set of objectives and plans focusing on cost reduction and quality improvement. They are aware that the path to success and improved profits will not be easy and will require flexibility and willingness to change from all levels of the organization.

NZ manufacturers are increasingly aware that they have to operate in an international market. The removal of quantitative restrictions on manufactured imports has increased the exposure of NZ manufacturers to foreign competition. This process will continue as tariff barriers are reduced or eliminated. Also, domestic consumers have raised their expectations of what they require in NZ-produced goods now that they have seen the quality of finish or design or the functionality of many newly imported products.

New Zealand has suffered a decline in its share of world trade over the last few years. Raw and semiprocessed goods are a declining proportion of world trade while manufactured goods (elaborately transformed manufactures) are increasing their share. The extension of the Australia–New Zealand Closer Economic Relations Trade Agreement (ANZCERTA), as well as other changes that have freed up trade into New Zealand have increased the pressure on manufacturers here.

As a result, manufacturers are pushing ahead with more export sales and expanding offshore production. Forty percent of NZ manufacturers are planning to increase offshore production, which must reflect an attempt to overcome the weakness in their ability to compete on price.

While New Zealand manufacturers may have had a tough time over the last few years, there is overwhelming determination to continue in business. When asked to describe the basic thrust of their business unit over the next two years, no one was prepared to withdraw from markets or exit the business. Only one respondent considered sacrificing market share. Building market share is the most favored strategy. There was a statistically significant difference between the industry groups on this matter. The machinery and industrials were strongly inclined to the build strategy, while a smaller majority of the consumer goods group are aiming to build market share. For the other two groups, electronics and basic industries, the majority of respondents are planning a hold market share strategy into the mid-1990s.

Competitive Priorities

The importance of dealing with the New Zealanders' perceived weakness in competing on price and the increasing demands for high quality lends a sense of urgency to the actions they plan. However, the current profitability crisis means many firms are finding it difficult to invest in research and development and new production technologies. NZ manufacturers are also vulnerable on the competitive dimensions of dependable delivery, fast delivery, high-performance products, and introducing new products. Activities to deal with these weaknesses have received relatively low emphasis in the plans of respondents.

In order to build market share while coping with the inability to make profits in price-competitive markets, manufacturers attach the highest importance to reducing costs by improving direct labor productivity, reducing overhead and materials costs, reducing inventories, and improving conformance quality (reducing defects). They plan to do this by linking their manufacturing strategy to business strategy, with increased efforts to train staff, particularly supervisors, and through quality function deployment. NZ manufacturers also see advantages in developing new processes for existing products, better integration of information systems, improving inventory control systems, and giving workers more tasks and responsibility, and plan to aggressively employ these tactics in coming years.

Performance

Performance indicators that have not improved since 1987 are those that have close links with the areas of weakness found in the section on competitive abilities—unit production costs, manufacturing lead time, procurement lead time, and delivery lead time. Quality as perceived by the customer has increased little compared with what one might have expected given manufacturers' perceptions of the importance of quality and their degree of strength. Also compared with the increases in quality being achieved by international competitors, NZ manufacturers need to work harder and smarter in this area.

The most important objectives for the manufacturing function are reducing costs and improving quality. Considerable emphasis on training and skills development at all levels in the organization will be an important feature of New Zealand manufacturing in the future.

Over the last three years, the typical respondent has found it difficult to make much improvement on production costs and lead times. The importance of quality and quality improvement have become critical, and while many manufacturers have accomplished much in this area over the last few years, it will be some time at current rates of improvement before they become competitive in world markets.

The moves that the government has made since 1985 to free up the economy have had a massive effect on the manufacturing sector. It appears that manufacturers accept deregulation and want to build market share but find the frequent changes of policy or timing not conducive to providing a settled environment in which to set business plans. One respondent remarked on "the difficulty of operating a long-term export-oriented business in a turbulent domestic environment."

Staff attitudes were seen as one the critical barriers to overcome in the 1990s. Good employee communication policies have paid off for many manufacturers in the recent past, but there remain significant differences between the industry groups on how far down the organization the strategies and plans of the business unit are known and understood.

Action Plans

NZ manufacturers believe that further increases in direct labor productivity are possible and ranked this as their most important objective in the next two years. Trimming waste and increasing efficiency were overwhelming concerns among NZ manufacturers. Their objectives include reducing unit costs, overhead and materials costs, as well as inventories. A small increase in labor costs between 1990 and 1992 after a decrease between 1989 and 1990 will add to the challenge of reducing costs of production. Increased attention will therefore need to be paid to programs that reduce waste and improve efficiency of materials use. These objectives compare favorably with the perceived weaknesses NZ manufacturers reported.

SINGAPORE: SOUTHEAST ASIA'S SOLID PERFORMER[5]

Over the last 10 years, Singapore has attracted a number of international manufacturing firms who operate there in conjunction with local firms or employees. Manufacturing industry is dominated by electronics manufacturing, which takes

---

[5]This section has been adapted from Professors Ta Huu Phuong and Lee Soo Ann of the National University of Singapore, "The Singapore Manufacturing Futures Survey–1990 Executive Summary," with their permission.

advantage of the highly educated and skilled workforce. The data from the Singapore Manufacturing Futures Project reflect these facts: The proportion of firms managed by Singaporeans was 37 percent, while 63 percent were managed jointly by local and foreign citizens; the majority are electronics firms.

Manufacturing industry has enjoyed good business through the 1980s, with continued upward trends in growth and profitability projected for the 1990s. More than 97 percent of the responding firms reported making profits. The net pre-tax profit figures ranged from 5 to 20 percent, with about 5 percent of firms reporting a profit of more than 20 percent. The year 1990 also saw a continued increase in manufacturing capacity utilization. About 47 percent of the responding firms operated above the 90 percent capacity level, while one fifth of them reported full capacity. The favorable economic climate was further reflected in sales growth. Almost all companies recorded positive average annual growth in sales volume. During the year, 31 percent of the firms grew by over 20 percent, while 61 percent have growth of less than 20 percent.

On the list of current problems faced by the manufacturing unit, the number one worry centered around high business expenses, particularly the cost of materials and the cost of labor. The managers in the survey indicated that they would emphasize the reduction of unit cost between 1990–92. Increasingly, manufacturers feel the strains caused by labor shortages, particularly for production workers. Future shortages may erode the country's competitiveness vis-à-vis other newly industrializing economies.

Competitive Priorities and Strategic Direction

Singapore manufacturers are keenly aware that they must offer quality and reliable products and provide timely deliveries to their customers in order to compete successfully in today's global markets. These manufacturers also perceived themselves to be most competitive on the non-cost factors of quality, service, and dependable deliveries. Indeed, all firms reported significantly improved performance in overall quality as perceived by customers from 1988–90, with three quarters of them reporting improvement in customer service and product variety.

Looking ahead to the next two years, Singapore manufacturers plan to increase market shares and to enter new markets by putting more effort into linking up manufacturing strategy to business strategy, and by placing more emphasis on quality function deployment (QFD) and training of their workers. The implementation of advanced information systems in manufacturing is also a high priority plan.

Quality, delivery, and service continue to receive great attention from Singapore-based management. These abilities were considered most important to enable firms to stay competitive in the marketplace over the next five years. In particular, the ability to offer consistent quality and provide reliable products are deemed to be of greatest importance.

Performance Improvement

Using 1988 as a base year to measure performance in various manufacturing activities, all the responding firms reported improved performance in overall quality as perceived by their customers. About three quarters of them reported performance improvement in customer service, market share, and product variety.

Action Plans

In terms of the importance of manufacturing objectives in the next two years, improvement in quality conformance, direct labor productivity, and delivery reliability were considered as most critical.

Linking manufacturing strategy to business strategy will be given the most emphasis by the firms over the next two years, followed by the training of workers, supervisors, and managers and the implementation of information systems in manufacturing.

Quality standards, cost of materials, and availability of production workers were the top major concerns of the surveyed firms. In general, major concerns are related to cost and quality.

BENCHMARKING THE PACIFIC RIM

The responses of manufacturers in these diverse economies dotted around the Pacific Rim reflect the importance of benchmarking their capabilities. We see a number of common factors that provide great challenges. First, these countries are no longer immune to the impact of global competition; the new economic models being developed in Australia, Mexico, and New Zealand reflect the changing government policies over the second half of the 1980s toward more open markets and the phasing out of the import substitution philosophy.

Second, the need for bigger markets has pushed some countries into pursuing trade agreements such as the United States-Canada-Mexico Zone and the Australia–New Zealand Free Trade Area. Korea and Singapore have not followed this line as yet, although regional initiatives do exist involving them. Third, with the exception of Singapore with its longer experience of free trade, many manufacturers in the Pacific Rim are experiencing a significant profit squeeze.

For managers in Pacific Rim manufacturing firms, this means that there is great pressure to improve quality and new product development processes that yield high-performance products. It is only by improving these aspects of their operations that manufacturing managers in the Pacific Rim will achieve their goals of developing products for niche markets. This calls for more flexibility in operations, tighter links between design and production, and an increased emphasis on training and gaining leverage from investments in information systems.

One of the important lessons learned in the 1980s in these countries was the significance of good communication policies and employee involvement in the change process to overall manufacturing performance. Labor unrest is of concern in Korea, but in New Zealand and Australia, initiatives by the nations and legislation by the governments have reduced the potential for unions to be a disruptive element. The Pacific Rim manufacturers are running hard to catch up with the biggest force in the area—Japan—in terms of quality and products. There is much ground to be made up.

The overview of the strategic manufacturing practice in these five countries should help the reader to understand how the Strategic Benchmarking Questionnaire and the results of the Futures Survey in the three main industrialized regions in the world can be used to evaluate manufacturing activities outside these regions. The patterns we described for Europe, the United States, or Japan are not the ultimate for manufacturing. Other development patterns are feasible and necessary.

There are large differences between the manufacturing strategies of companies in mature economies based on natural resources, such as Australia and New Zealand; those of a Korea trying to adapt its economy from that of a developing country to that of a fully industrialized nation; those of a Mexico discovering free trade and a freer economy; and those of a Singapore that defines her role as a haven of free economic trade and as a communication node in the global economy. Manufacturers in each country have developed different strategies, emphasized different action plans, and made different improvements in performance. No wonder! They have different role models and have to adapt to different trade partners.

What the short descriptions of manufacturing in each Pacific Rim country in this chapter illustrate is how each country, starting from the specific economic situation in which it works, can develop a coherent set of strategic directions, competitive priorities, and action plans. At each stage, it is valuable to relate one's own choices to the patterns and choices made by role models or trading partners, in order to check for complementary actions or potential conflicts.

CHAPTER 9

FACTORIES OF THE FUTURE*

The factories of the future envisioned by manufacturing executives in the large successful firms based in Europe, America, and Japan are different from each other not only in terms of what they are designed to do, but also in terms of the key tasks required to create them. Supported by significant improvements in the past, and faced with rapidly changing worldwide business environments, these companies are looking into the future with newly shaped competitive priorities and visions.

Dominant visions of the factory of the future are emerging in each region. Manufacturers in the United States, Japan, and Europe are all pursuing a more integrated factory that can yield higher quality goods. However, each of these factories of the future is being designed to achieve different objectives. U.S. manufacturers are moving toward the *value* factory of the future, designed to build high-quality products at a reasonable price. European companies are moving toward the *borderless* factory of the future, equipped to do business in the newly forming pan-European market as well as its newly forming cross-functional organizations. In Japan, manufacturers are working steadily to build the *design* factory of the future, oriented toward producing not just products in volume, but a

---

*This chapter has been adapted from Professors Jeffrey G. Miller and Jay S. Kim of Boston University, Arnoud De Meyer and Kasra Ferdows of INSEAD, and Jinichiro Nakane and Seiji Kurosu of Waseda University, "Factories of the Future," with their permission.

TABLE 9–1
Top Five Competitive Priorities in the Next Five Years

Europe	Japan	United States
Conformance quality	Product reliability	Conformance quality
On-time delivery	On-time delivery	On-time delivery
Product reliability	Fast design change	Product reliability
Performance quality	Conformance quality	Performance quality
Delivery speed	Product customization	Price

continuing stream of new designs that are highly customized for individual use.

In Table 9–1, a comparison of the competitive priorities for each region reveals the Japanese on a path distinct from the strategic paths of manufacturers in the United States and Europe. The similarities between Americans and Europeans are striking although not new. We have observed this similarity in our research every year since 1983. The top five priorities from the two regions include providing consistently low defect rates, dependable deliveries, product reliability, and high-performance products (products with advanced features and functionality). Defect rates, reliability, and performance are three key dimensions of what is commonly viewed as quality. The importance of high-performance products to both the Americans and Europeans, but not the Japanese, indicates the Western focus on advanced product technologies to provide the basis for superiority in product performance. The Americans are distinguishable from the Europeans by their greater emphasis on the ability to compete on price. Thus, we conclude that Americans are focused on competing more on the basis of price-adjusted value. In contrast, the priorities of the European manufacturers reflect a focus on quality and delivery to their customers (in a Europe without borders).

More striking differences are found in the comparison between the Japanese manufacturers and their counterparts in the other two regions. While Japanese firms also place a strong emphasis on quality and delivery-related priorities, they assign much higher priorities to the ability to make rapid design changes in highly customized products. Hence, the design factory, built to be flexible enough to produce an ever-changing stream of products. In addition, the Japanese focus more heavily on reliability and conformance as aspects of quality and less on product performance. This is consistent with the Japanese emphasis on process technology and simplicity, as reported by many other researchers.

In Table 9–2, we notice a remarkable difference in the investments being made to improve manufacturing by the U.S. manufacturers compared to their competitors in the other two regions. Improvement programs in the United States are

TABLE 9–2
Most Important Improvement Programs in the Next Two Years

United States

1. Linking manufacturing strategy to business strategy
2. Giving workers broader tasks and more responsibilities
3. Statistical process control
4. Worker and supervisor training
5. Interfunctional work teams

Europe

1. Linking manufacturing strategy to business strategy
2. Integrating information systems in manufacturing
3. Quality function deployment
4. Training of supervisors, workers, and managers
5. Integrating information systems across functions

Japan

1. Integration of information systems in manufacturing and across functions
2. Developing new processes for new products
3. Production and inventory control systems
4. Developing new processes for old products
5. Linking manufacturing strategy to business strategy

heavily focused on workers and teams of workers. American manufacturers are eagerly accepting the idea that the greatest advances in quality and cost will come by positioning the workforce to make a greater contribution to improvement and by breaking functional boundaries. Our surveys also show that U.S. firms carry the largest percentage of their costs as overhead among the three regions. By reallocating responsibilities to individuals lower in the organizational pyramid, and breaking bureaucracies with direct interfunctional teams, the Americans can improve their ability to create the value factory of the future at a lower cost than before.

In contrast, the Japanese manufacturers place a strong emphasis on the development and improvement of fundamental process technologies. Along with information systems integration, this strong orientation toward process development supports their thrust toward factories that can make reliable, rapid design changes in customized products. The Japanese design factory of the future is fundamentally based on new processes that can overcome the enormous complexity involved in delivering large numbers of new and custom products with high quality in a short lead time.

European manufacturers are investing in a variety of programs and activities that hint at the major restructuring their organizations are undergoing to adapt to the newly forming pan-European market. Restructuring implies a simultaneous change in the culture, coordination techniques, and configuration of a firm. Europe's investment in integrated information systems is designed to enhance its ability to coordinate its activities across borders. Quality function deployment enhances a firm's ability to design products that are acceptable to the new European market. Training is necessary to build new skills and a new culture for competition in the pan-European vision of the borderless factory of the future.

Whether they are based on a vision of a value factory, a design factory, or a borderless factory, the factories of the future in each region are all focused on the development of organizations in which manufacturing strategy is tightly linked with business strategy. All three regions report a high emphasis on the linkage between manufacturing's functional strategy and overall business direction. This result signifies recognition of the manufacturing function's strategic importance in the increasingly competitive global marketplace.

CHANGING PERFORMANCE BENCHMARKS

While each emerging vision of the factory of the future is designed to perform well on a particular set of dimensions, no competitor can afford to perform on any dimension at an unacceptable level. In this race, all aspects of performance are important. For this reason, performance benchmarks have become increasingly important to manufacturers around the globe. Overall, large Japanese manufacturers perceive themselves to be the performance leaders on many key measures; U.S. manufacturers have been fighting back and gaining ground; and European performance has begun to improve dramatically.

Statistical analysis shows few differences in perceived competitiveness across regions on such critical variables as conformance quality, product reliability, cost, and delivery dependability. However, regional differences do show up in other dimensions, as seen in Table 9–3. European and American companies see themselves as leaders in designing high-performance products. American firms also believe they have significant advantages in their ability to distribute products across a broad range of channels and customers. Japan considers itself a leader in flexible manufacturing; that is, making rapid design changes, introducing new products quickly, and handling volume surges.

Not one of the region's leading manufacturers believe they dominate on more than one or two performance criteria, underscoring the importance of rates of improvement in the competitive race through the 1990s. The manufacturers in our sample reported significant improvements in several performance measures in 1988 and 1989. The average annual rate of improvement

TABLE 9–3
1990 Competitiveness Index

	United States	Japan	Europe
Price	3.98	4.11	4.14
Design change	3.89	4.32*	4.25
Volume change	4.31	4.61*	4.34
Product mix change	4.43	4.62	4.41
Broad product line	4.59	4.44	4.60
Conformance quality	4.94	4.91	5.03
Performance quality	4.92*	4.43	5.01*
Product reliability	5.23	5.02	5.17
Delivery speed	4.61	4.46	4.52
On-time delivery	4.82	4.79	4.67
After-sales services	4.86	4.62	4.67
Product support	4.85	4.59	4.67
Broad distribution	4.81*	4.51	4.49
Product customization	4.56	4.70	4.63

Note: The table shows the normalized mean scores of the respondents from each region on a question asking them to rate their perception of their capabilities on each performance measure in comparison to their competitors. A 7-point scale was used, where 1 was weak and 7 was strong. All respondents have a tendency to overrate their competitiveness. Therefore, we have normalized the scores by each region's tendency to overrate.
*Asterisks indicate scores that are significantly different from other regions.

on eight critical performance dimensions is shown in Figure 9–1, and Table 9–4 represents the complete list.

The most notable feature of Figure 9–1 is the overall high rate of improvement for Europe. In past surveys (1985–87 time frame), Europe's rates of improvement were significantly lower than those for the United States and Japan. During the mid-80s, quality, on-time deliveries, and delivery speed were improving at a lackluster 2.5 percent per year among the European firms. In contrast, the rate of quality improvement in Europe during 1988–89 nearly tripled to 6.1 percent per year, suggesting that European firms have finally joined Japanese and American competitors in the quality revolution that has been sweeping the industrial world. Rates of improvement in delivery reliability and speed also increased concomitantly.

FIGURE 9–1
Performance Improvements, 1988–1989

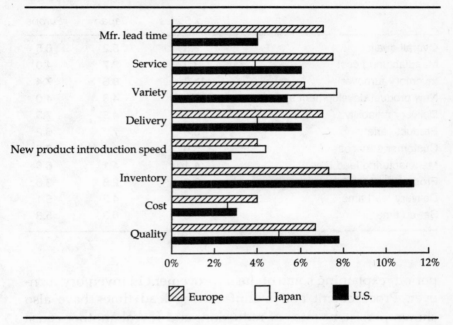

The improvement we observe in the key variables is the result of catch-up efforts on the part of some companies, and major advances on the frontiers for others. The benchmarks for the following four areas are the fastest changing and so deserve special attention.

Inventory Turnover

All three regions report significant improvements in inventory turnover. American firms improved at an annual rate of 11.3 percent per year, the Japanese at 8.3 percent, and the Europeans at 7.4 percent. There is no statistical difference between the rates of the Europeans and the Japanese, but the U.S. rate was significantly faster than the others, as it was in 1985–87.

Table 9–4 shows that set-up time has been reduced by about 6 percent per year in each region over the 1988–89 time

TABLE 9–4
Annual Rates (percent) of Improvement for 1988–1989

	United States	Japan	Europe
Overall quality	7.8	5.2	6.5
Manufacturing cost	3.0	2.7	4.0
Inventory turnover	11.3	8.3	7.4
New product development speed	2.8	4.3	4.0
Delivery reliability	6.2	4.3	7.3
Product variety	5.6	7.7	6.2
Customer services	5.9	4.0	7.2
Manufacturing lead time	4.1	4.1	6.6
Procurement lead time	2.3	2.8	3.8
Delivery lead time	2.4	4.3	5.1
Set-up time	5.9	6.3	5.8

period, explaining some of the improvement in inventory turn-over. Procurement and manufacturing lead times have also shown persistent rates of reduction, and reveal another cause for inventory reduction. The rates of decline in manufacturing lead time for European manufacturers were significantly greater than those of the Americans and the Japanese.

Quality and Customer Service

The relentless drive towards better quality continues around the world. We have seen in previous surveys that substantial attention has been paid to quality improvements over the last eight years. This trend continues in Japan and the United States, with signs that Europe has embraced it as well. The Americans report 7.8 percent annual rates of improvement in overall quality, while European and Japanese firms improved quality by 6.5 percent and 5.2 percent, respectively. More than ever, product quality has become a given in industrial competition, qualifying the players in the game rather than separating the winners and losers.

A firm's perception of its performance in the area of customer service was measured for the first time in the 1990 survey. The high rates of improvement in customer service show that increasing attention is being paid to non-product quality as well; it is the area of quality in which the Europeans have shown the most improvement, with the Americans close behind.

Time-Based Competition

Time continues to be a key element of competition for all three regions, as shown by the high rates of improvement in delivery reliability and speed. The Americans and Europeans report 6–7 percent annual rates of improvement in the reliability of delivery; the Japanese report 4.3 percent per year. Table 9–4 shows that the Europeans and the Japanese are improving delivery speed at faster rates than their counterparts in the United States.

As mentioned before, this improvement in delivery speed and reliability has not been accomplished at the expense of higher inventories. This suggests that the manufacturers in all regions, particularly in Europe and the United States, have been increasingly successful in their efforts to overcome the inherent variability and uncertainty in the order-filling process. Table 9–4 provides further clues on this critical competitive dimension: manufacturing lead times have been reduced at an annual rate of 6.6 percent in Europe, and 4.1 percent in America and Japan, as well. At the same time, delivery and procurement lead times were shortened considerably.

Reduction in the time required for new product development has received much attention recently. The rates of improvement reported by our sample of manufacturers are fairly modest—4.3 percent by the Japanese, 4.0 percent by the Europeans, and 2.8 percent by the Americans. The Japanese exhibited a similar rate of improvement on this dimension over the 1985–87 time period, according to our previous surveys. Given the long cycle times associated with many new product introductions in the United States and Europe, it will be several years before substantial reductions can be realized.

Flexibility

As seen in Table 9–3, Japanese manufacturers excel in their ability to change product designs and respond to changes in volume or product mix. Along with their speed at introducing new products, the Japanese are also proving themselves more flexible to produce a broad variety of products; Figure 9–1 shows that the Japanese are substantially ahead of the Europeans and the Americans in terms of their rate of improvement on this variable as well. Flexibility provides competitors with the ability to respond better to individual customer needs. Therefore, greater flexibility is an advantage for the competitor with globally diverse operations. Flexibility also allows manufacturers to serve local markets with a wide variety of individually tailored products.

It is notable that the fastest rates of improvement in each region have occurred on the measures that are most closely associated with each region's vision of its factory of the future. The United States has improved the fastest on the two dimensions that add to its ability to deliver value: quality and inventory. The Europeans have improved the most on those dimensions that allow them to serve broader pan-European markets better: delivery reliability, customer service, and shorter lead times. The Japanese have improved the fastest in their ability to change product designs and deliver a broader variety of products from their factories.

EFFECTIVENESS OF IMPROVEMENT PROGRAMS

How did the manufacturers in each region achieve the improvements in performance they reported? Table 9–5 shows the improvement programs with the highest and lowest payoffs in each region.

TABLE 9–5
Pay-Off from Activities, 1988–1989

Europe

	1. Training of supervisors, management, and workers
Highest	2. Manufacturing reorganization
Pay-off	3. Linking manufacturing to business strategy
	4. Quality function deployment
	5. Developing new processes for new products

⋮

	22. Quality circles
Lowest	23. Activity-based costing
Pay-off	24. Design for manufacture
	25. Closing/relocating plants
	26. Robotics

United States

	1. Interfunctional work teams
Highest	2. Manufacturing reorganization
Pay-off	3. Statistical process control
	4. Linking manufacturing to business strategy
	5. Just-in-time

⋮

	22. Flexible manufacturing system
Lowest	23. Integrating information systems across functions
Pay-off	24. Integrating information systems in manufacturing
	25. Activity-based costing
	26. Robotics

Japan

	1. Developing new processes for old products
Highest	2. Developing new processes for new products
Pay-off	3. Quality circles
	4. Computer-aided design
	5. Quality function deployment

⋮

	22. Computer-aided manufacturing/flexible manufacturing system
Lowest	23. Integrating information systems across functions
Pay-off	24. Hiring new skills from outside
	25. Activity-based costing
	26. Closing/relocating plants

171

Though manufacturers in the three regions have grown towards each other in performance, the way they did this was quite different. The similarities and differences are intriguing. In broad terms, the Americans and the Europeans have benefited from structural changes in manufacturing—reorganization of the manufacturing function, redeploying workers, and linking manufacturing more closely to the business strategy. The Japanese, on the other hand, seem to find the most benefit from fundamental process development, as well as in quality function deployment and quality circles.

Training of workers, supervisors, and managers has paid off handsomely everywhere—making it to the top of the list for the Europeans, and in the top tier of the American and Japanese lists. For the Americans, the investments in statistical process control and just-in-time programs are starting to pay off. The Japanese continue to benefit from quality circles (but, interestingly, not the Europeans) and from the constant search for new production processes for old and new products.

By comparing the lists of high-pay-off items in the 1990 survey with those of previous surveys, we find at least one relatively new item on each of the three lists. In the American list, it is interfunctional work teams. The shorter lines of communications, reduced bureaucracy, freer flow of information, mutual trust, and other benefits from this approach are evidently proving to be worthwhile. On the European list, the new item is quality function deployment (QFD): a set of techniques for determining and communicating customer needs and translating them into product and service design specifications and manufacturing methods. QFD is also on the Japanese list of high-pay-off programs.

For the Japanese, a new item on the high-pay-off list that deserves attention is computer-aided design (CAD). It is interesting that CAD has started to pay off, while other computerization programs such as computer-aided manufacturing (CAM), integration of information systems between manufacturing and other functions, and even flexible manufacturing systems (FMS) are all in the bottom of the pay-off list in Table

9–5. The focus on CAD is symptomatic of the Japanese focus on the design factory of the future.

One surprising result was the low-payoff showing for activity-based costing (ABC) in all three regions. Many manufacturers in our sample had not started an ABC program. However, even among those who had, the response was poor. We believe that the low-payoff scores for ABC reflect a lack of fundamental yet important changes in some of the most entrenched control practices in companies attempting it. Such changes are difficult and do take time. Moreover, we do not believe that ABC alone has the potential for a positive impact of the same magnitude as other more fundamental investments. ABC may be helpful in steering and measuring the course of continuous improvement, but it cannot provide the punch and impact that other quality and process improvement programs can provide.

Finally, we note the low pay-off given to efforts to improve design for manufacture (DFM). The Europeans put it near the bottom of the list, and the Americans and the Japanese in the lower third. We believe that while the focus of DFM (improving engineering-manufacturing communication) is fundamentally right, it is losing ground in popularity because it places the focus of design improvements on manufacturing and engineering issues and not the customer's needs. Quality function deployment achieves many of the same aims as DFM, but puts the emphasis on integrating the customer, marketing, *and* engineering and manufacturing.

Behind the programs for improving manufacturing effectiveness is a fundamental issue of strategy implementation. Who in the organization has knowledge of its goals and plans and is called on to deliver them? Figure 9–2 shows that Japanese and U.S. firms have the widest involvement of their employees, and that European firms remain the most hierarchical in their management approach. Employee participation has not been a hallmark of American manufacturing for long, however. The 1984 U.S. survey showed that only 2 percent of the firms in the sample made all employees aware of

FIGURE 9–2
Respondents (percent) Who Share Knowledge of Goals and Plans, 1990

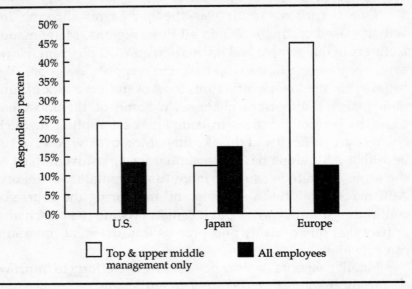

goals and plans. In 1990, nearly 14 percent of U.S. respondents reported awareness on the part of all employees. Detailed examination of the data shows that in those firms where all employees have not been made aware of goals and plans, there is a greater tendency to share this information with middle management.

FACTORIES OF THE FUTURE AND THE GLOBAL MANUFACTURING RACE

The Americans are building a value factory of the future, the Japanese a design factory of the future, and the Europeans a borderless factory of the future. The strategic intent of the majority of manufacturers in each region is clear. But why does each vision make sense for one region and not for another? What are the risks inherent in their pursuit?

The argument for the Japanese factory of the future comes from increasing pressure in Japanese domestic and offshore markets. Japanese firms operate in a country whose domestic customers are demanding more sophisticated new products. At the same time, the high value of the yen to other foreign currency makes the Japanese firms less able to export low-valued products or to compete on price. This combination of forces creates an environment in which the ability to rapidly introduce, change, or modify new products becomes a competitive necessity. With a design factory, the Japanese firm can meet the explosive growth in its domestic demand for newer, more sophisticated, and more individualized products, and compete for offshore markets by providing a continuing stream of highly differentiated products that are less susceptible to price competition. It also allows the Japanese to exploit one of their national assets: a large pool of well-trained and (comparatively) low-cost engineers.

In contrast to Japan, demographic and economic forces predict relatively low rates of consumption growth in the United States. However, improved quality and a favorable exchange rate offer American producers an opening to compete globally on the basis of value; that is, good quality for the price. At the same time, the United States remains (and will remain even after 1992) the largest homogeneous market in the world, offering advantages to manufacturers who can exploit economies of scale. Though product development is also important for the Americans, especially in high-tech industries such as electronics, it does not provide a fundamental driving force across industries as it does in Japan. An emphasis on the value factory of the future calls for actions that will simultaneously increase quality and reduce costs, which is what we see the Americans doing. This strategy also exploits a previously untapped resource, the American production worker. Historically, the American factory worker has not been required to think about ways to improve quality and reduce costs. Those tasks have been performed by more highly paid "overhead" employees who design and direct factory workers' activities. Use of these lower cost workers as a source of continuous improvement is a relatively inexpensive way for U.S. firms to deliver more value.

The borderless factory of the Europeans responds to their needs to adapt to the new Europe of 1992. They are driven not so much by demographics and exchange rates as they are by new political and market realities. With the formation of the EC, Europeans are responding to many more fundamental and enduring changes than the Japanese and Americans. The massive restructuring of information systems, organizations, and corporate cultures leads to a factory of the future that is designed to do the same things as the factory of old—excel at quality and delivery. But the borderless factory will be structured to deliver on a pan-European rather than a country-by-country basis.

Some evidence for these interpretations, as well as further insights into the strategies, is provided in the following two figures. Figure 9–3 shows that manufacturers in each region have aggressive intentions with respect to market share. The high percentage of American and European manufacturers'

FIGURE 9–3
Future Market Strategies

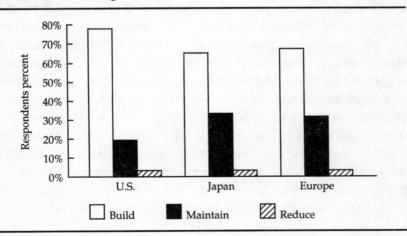

intent on building market share is consistent with the American's opportunity to grow in export markets and the European's opportunity to build new pan-European markets. We believe that the large percentage of Japanese firms with a Maintain posture reflects two factors. The first is the pressure put on these firms to keep up with the high growth rate of the Japanese domestic economy, and the second is the impact of trade pressure from America and Europe.

Figure 9–4 shows the planned rate of increase in the offshore portion of sales, production, and purchasing by producers in the sample. It shows the Americans are following an aggressive program of exports that is already appearing in economic statistics, and this aggressiveness is consistent with the priorities we have seen. We believe the increases in offshore production and purchasing by the Americans reflect their intention to become more global, to produce locally in the offshore markets, and to balance procurement patterns in accordance with local requirements and countertrade issues.

FIGURE 9–4
Changes in Offshore Activities, 1989–1992

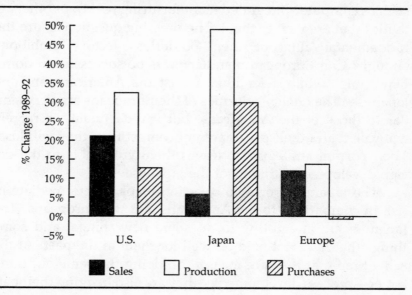

The Europeans are substantially less aggressive in their plans to export from the EC and to produce offshore. We believe that this is an indication of their focus on changes in Western Europe in anticipation of a unified market by 1992, as well as the challenge posed by the opening of Eastern European markets. The Japanese are reacting to trade pressure by increasing offshore production, with a consequent reduction in the growth rate of exports.

The disadvantages and risks associated with the pursuit of these factories of the future are significant. The Americans are following a strategy of "doing better what we already do." An analysis of their action plans shows that the top five priority items are limited to activities that are also on the high-pay-off list. The question for the Americans is whether they can and should move beyond the quality revolution as the Japanese have in order to meet the competitive challenge presented by the Japanese and adapt to the opportunities of the 21st century.

The Europeans are clearly building a factory of the future that is capable of "doing new things" in a unified European market. They are taking the most risks by investing in those activities that have had either low pay-off for them historically, such as computer-integrated manufacturing (CIM), or other activities that are new to them. The pressing question is, are the fundamental aims of their borderless factory ambitious enough? Can European manufacturers be successful in doing new things while defending against the Americans and the Japanese? The quality priorities of the Europeans are very similar to those of the Americans, but if their quality is merely equivalent and their prices are not competitive, they will lose. Their current strategy also leaves them vulnerable to the design/development strategy of the Japanese.

The Japanese portfolio of action plans is intermediate in risk in comparison to the Americans and the Europeans. The Japanese are attempting to do some new things, and some things that have not paid off well for them in the past. At the same time that they are focused on doing the same old thing better—process development—they are also hedging their bets by moving large parts of their industrial base to America and Europe, while they learn how to do the new things at home. This move, as we have noted earlier, poses significant challenges for the Japanese as they learn to work with and manage foreign workers.

Each factory of the future represents a viable option for competitors within the region in the short term. However, there are long-term risks associated with adopting each version. Moreover, we see that competitors holding separate visions will influence each other over time as they already have in the previous decade. The end result at some currently unknown time in the future may be a unified vision of the factory of the future.

PART 3

APPENDIXES

SURVEY METHODOLOGY

The Manufacturing Futures Survey was initially developed at Boston University in 1981 by Professor Jeffrey Miller and administered in North America in the Fall of that year. The survey results were released in the Spring of 1982 and became known as the 1982 Manufacturing Futures Survey. The survey is now administered by an international team of researchers from the Boston University School of Management, INSEAD in Fontainebleau, France, and Waseda University in Tokyo, Japan. Executive summaries from each year's survey have reported the general findings. As of 1990, the survey will be conducted every other year. It is currently administered in 18 nations, including the Western European countries, Japan, Korea, Australia, New Zealand, Mexico, and the United States.

The 1990 U.S. Manufacturing Futures Survey is a nine-page questionnaire with over 200 individual questions. It forms the basis for the Strategic Benchmarking Questionnaire included in this book at the end of Chapter 3. The core questions in the survey are divided into two sections of roughly equal length. The first section focuses on obtaining a manufacturing business unit and product profile. Included are questions that allow the company to be categorized by industry, product, and market type. Overall financial, performance, and strategy profiles are gathered here. The first section also contains questions on manufacturing cost structure, staffing, historical manufacturing performance, and offshore activities. The second section is devoted to gathering data on the respondent's intentions for improving manufacturing effectiveness. Questions included in this section are manufacturing's objectives, past action programs and their pay-offs, and future plans for strategic actions.

The non-core questions included in the 1990 survey were designed to gain insight on how various functional decisions are linked to form a manufacturing strategy. The degree of linkage between product and process decisions was explored in particular, along with questions regarding concurrent engineering. The 1990 survey also included five open-ended questions regarding the lessons from the 1980s and the leadership challenges that are foreseen by the manufacturing executives. Chapter 3 describes these sections and the questions in more detail.

Respondents and Sampling Frame

U.S. Respondents

In the United States, the 1990 Manufacturing Futures Survey was sent by first-class mail with a personally addressed letter to 1,306 potential respondents over the period of January 15 to 30, 1990. The potential respondents were drawn from three different panels: A, B, and C. The A and B panels together accounted for 884 names. These names and addresses were initially obtained from the 1982 Fortune 500 listing of the largest industrial corporations and supplemented by lists of senior manufacturing participants in various executive programs. Panel A (434 members) included those who had responded to the survey at least once in 1984, 1985, 1986, 1987, or 1988. The response rate from this important longitudinal group was 23.7 percent. The response rate for the B panel (450 members) was 12.7 percent. The B panel included previous non-respondents as well as the names of individuals who had attended various executive programs that were focused on manufacturing strategy. The C panel (422 members) was proportionately drawn, by two-digit SIC (standard industrial classification) code, from a commercially available list of leading U.S. firms by industry (Dun's Business Rankings, 1986). The response rate for this panel was 5.7 percent.

A follow-up letter was sent to those who had not responded by the end of February. A survey hotline was maintained to answer by phone call or mail any questions respondents had about the survey. Each non-respondent was reminded by another follow-up letter in March. By April 30, 184 valid responses had been encoded. Two responses were eliminated because they were duplicates. The analysis in this monograph is based on 182 responses.

The 182 participants in the 1990 U.S. Manufacturing Futures Survey are top manufacturing executives in business units that operate in five industry categories. The conclusions in this report are based on an analysis of the 182 usable surveys received by April 30, 1990. The typical respondent is the vice president of manufacturing for a division with sales of $190 million, has been with the company for 17 years, and has held their current position for four years. Our analysis has shown that the survey respondents are typical of large American manufacturers in most respects. The most notable bias is the unusually large percentage of successful business units represented. The market share held by respondent business units averaged 39 percent, the 1989 unit growth rate was 7.6 percent, and the mean pre-tax return on sales was nearly 10 percent.

The distribution of U.S. responses by industry category was as follows:

Industry Category	U.S. Responses (percent)
Consumer packaged goods	17.6%
Industrial goods	21.4
Basic industries	13.2
Machinery	23.6
Electronics	24.2
	100.0%

Japanese Respondents

In Japan, the MFP Survey was initiated in May 1990, when questionnaires were mailed to 600 executives in large manufacturing business units in Japan. The survey was sent to a randomly selected list of executives in the manufacturing companies listed in the Japanese stock market. The executives who received the survey were asked to answer questions relating to current activities and plans from the perspective of their unit.

The cut-off date for the return of the questionnaires was the end of June 1990; 123 valid questionnaires were accepted. Respondents typically were high level managers or staff in manufacturing companies.

The following table shows the distribution of responding Japanese business units by industry category (major product). The greatest concentration of responding business units was in Electronics/ Electric and Machinery.

Industry Category	Japanese Responses (percent)
Consumer packaged goods	7.3%
Industrial goods	N/A

Industry Category	Japanese Responses (percent)
Basic industries	27.6
Machinery	38.2
Electronics	26.9
	100.0%

(concluded)

European Respondents

The methods used to administer the survey in Europe were similar to those employed in the United States and Japan. The following table describes the 223 respondents in terms of their industry and country distributions.

Industry Category	European Responses (percent)
Consumer packaged goods	20%
Industrial goods	9%
Basic industries	32%
Machinery	19%
Electronics	20%
	100%

Country of distribution	Number of Respondents
Austria	4
Belgium	23
Denmark	17
Finland	8
France	36
Germany	28
Great Britain	41

Country of distribution	Number of Respondents
Ireland	3
Italy	12
Netherlands	11
Portugal	2
Spain	12
Sweden	7
Switzerland	19
	223

(concluded)

The survey instrument was designed to be answered by a senior manufacturing executive of the responding business unit. A business unit was defined as "an entire company, a group, a division, or in some instances a plant, depending upon the organization of the company." A company may be the relevant business unit if the company was engaged in essentially a single product/line of business. A group or division may be more appropriate for a diversified operation. Plants are more relevant units of analysis where strategies vary significantly by location. Given the considerable natural variation that exists in corporate structures, our intent was to capture responses at a business unit level where manufacturing and market strategies were jointly determined.

Data Analysis

Surveys were directly encoded into a data file and uploaded to mainframe computers where SPSS (Statistical Package for the Social Sciences) statistical programs were maintained. Computer edits and consistency checks were used to detect and correct erroneously entered data. Inconsistent responses on surveys were followed up for clarification.

The SPSS statistical package was used for the analysis of the data described in this book. Respondent answers were used to categorize them according to membership within industry groups. Analysis of variance and the Scheffe method was used to identify those questions that differentiated between groups at the 0.05 level of significance or less. The overall F statistics and overall probability levels associated with each one-way analysis of variance are given in the appendixes.

Appendix Table Description

The tables in Appendixes A, B, and C contain basic descriptive statistical information developed from the responses to the Manufacturing Futures Survey in 1990. This information is intended to be used by individuals and firms attempting to strategically benchmark themselves against the leadership group of global manufacturers in the Manufacturing Futures Survey database.

Each of the table cells in Appendixes A, B, and C contains the mean response score or value for various reference groups for each of the questions indicated in the Strategic Benchmarking Questionnaire at the end of Chapter 3. Additional information contained in each table includes the standard deviation and the number of respondents in the reference group. The standard deviation is a measure of sampling error. The column labeled Total provides the overall mean response to each question. Furthermore, for each classification, the F statistic and the significance level (probability) derived from analysis of variance is indicated. A probability level less than or equal to 0.05 is considered statistically significant. In other words, it is expected that the difference between one or more of the sample reference group means in the table is due to an assignable cause. The appendix tables are described in more detail below:

Appendix A: Country/Regional Data

The reference groups that form the columns in this set of tables correspond to the largest countries or regions in which the Manufacturing Futures Survey was administered in 1990. The United States (U.S.) and Japan are relatively well defined and homogeneous. Europe, of course, is not a country but a region that can be defined by a number of cultural/ethnic/language borders. We have defined these regions in terms of the four typical categories:

Latin: France, Italy, Spain, and Portugal.

Germanic: Germany, Austria, Switzerland, the Netherlands, Belgium, and Luxembourg.

Anglo-Saxon: England, Scotland, Wales, and Ireland.

Nordic: Sweden, Norway, Denmark, and Finland.

Appendix B: Industry Data

The reference groups indicated by the columns in Appendix B are defined by the five industry groups. These industry groups are defined as follows:

Machinery

MBUs (manufacturing business units) in this category are involved in the manufacturing of finished machinery and equipment involving the fabrication and assembly of metal and electromechanical components.

SIC Code	Description
351	Engines and turbines
352	Farm and garden equipment
353	Construction, mining, and material equipment
354	Metal-working equipment
355	Special industrial machinery
356	General industrial machinery
358	Refrigeration and service machinery
359	Miscellaneous machinery
371	Motor vehicles
372	Aircraft
376	Guided missiles and space vehicles

Electronics

The products of the MBUs in this group all employ electronic technology extensively.

SIC Code	Description
357	Calculating machines and computers
361	Transmission and distribution equipment
362	Electronic industrial apparatus
363	Electronic household appliances
364	Lighting and wiring
366	Communication equipment
367	Electrical components

SIC Code	Description
381	Laboratory instruments
382	Measuring and controlling
383	Optical instruments
384	Medical instruments

(concluded)

Consumer Packaged Goods

The common denominator among business units in this group is that their products are all found on retail store shelves.

SIC Code	Description
201	Meat products
203	Canned fruits and vegetables
208	Beverages
209	Miscellaneous products
265	Consumer paper products
277	Greeting card publishing
283	Drugs
284	Cleaning, cosmetic, and toiletry
314	Footwear
342	Consumer hardware and hand tools
386	Photographic equipment and film

Industrial Goods

MBUs in this group manufacture products that are sold at the intermediate level in the industrial chain. They tend to purchase material from firms in the basic industry category.

SIC Code	Description
229	Miscellaneous textiles
249	Miscellaneous wood products
252	Office furniture
259	Miscellaneous furniture and fixtures

SIC Code	Description
264	Converted paper products
265	Paper containers and boxes
275	Commercial printing
306	Fabricated rubber products
341	Metal canes and shipping containers
342	Industrial hand tools
343	Plumbing and heating
344	Fabricated structural metal products
346	Metal forgings and stamps
348	Ordinance fabricated metal products

(concluded)

Basic Industries

MBUs in this category convert natural ores, metals, and organic compounds into basic materials. This group could also have been called the process industry group.

SIC Code	Description
221	Broad woven fabric
225	Knitting mills
262	Paper mills
281	Inorganic chemicals
282	Plastics, resins, and synthetics
283	Specialty inorganic chemicals
285	Paints and varnishes
291	Petroleum refining
301	Tires and innertubes
322	Glass and glassware
326	Pottery and related products
327	Concrete and gypsum products
329	Non-metallic mineral products
332	Iron and steel foundries
333	Nonferrous smelting and refining
335	Nonferrous rolling and drawing
336	Nonferrous foundries
339	Primary metal products

The reporting and classification methods used in Japan were somewhat different, resulting in the elimination of the industrial goods classification in this country. Most of the respondents in the industrial goods sector were reclassified in this country as basic industries.

Appendix C: Country by Industry Data

This set of tables provides information on how the five industry groups have responded on average to the survey in each of the three major industrial regions—Japan, the United States, and Europe.

Table Entries and the Strategic Benchmarking Questionnaire

The rows in each of the sets of tables in Appendixes A, B, and C correspond to the questions in the Strategic Benchmarking Questionnaire found at the end of Chapter 3, as indicated in the following table.

Question Type	Strategic Benchmarking Survey Question Number
Business unit growth strategy	A2
Selected business unit profile	A5
Importance of competitive abilities	A4(left)
Strength in competitive abilities	A4(right)
Manufacturing cost structure	A6, A7
Manufacturing performance improvement index	A9
Manufacturing objectives	B1
Past pay-off from programs and activities	B2(left)
Future emphasis on programs and activities	B2(right)
Internal environment	A3
Global plans	A8

APPENDIX A

COUNTRY DATA

Business Unit Growth Strategy by Country

Indicators	Country		
	Total	U.S.	Japan
Build market share			
Mean	0.69	0.79	0.63
Standard deviation	0.46	0.41	0.48
Number	549	182	143
Hold (defend) market share			
Mean	0.27	0.18	0.32
Standard deviation	0.44	0.38	0.47
Number	549	182	143
Harvest (maximize cash flow, sacrifice market share)			
Mean	0.03	0.03	0.04
Standard deviation	0.17	0.18	0.18
Number	549	182	143
Withdraw (prepare to exit the business)			
Mean	0.00	0.01	0.00
Standard deviation	0.06	0.07	0.00
Number	549	182	143

	Country			
	Europe			
Total Europe	Latin	Germanic	Anglo-Saxon	Nordic
0.66*	0.71	0.67	0.57	0.65
0.47	0.46	0.47	0.50	0.48
549	85	52	44	43
0.31*	0.26	0.31	0.41	0.30
0.46	0.44	0.47	0.50	0.46
224	85	52	44	43
0.02	0.01	0.02	0.02	0.05
0.15	0.11	0.14	0.15	0.21
224	85	52	44	43
0.00	0.01	0.00	0.00	0.00
0.07	0.11	0.00	0.00	0.00
224	85	52	44	43

*The ANOVA F-test calculated from the comparison between the United States, Japan, and Europe is significant at the .05 alpha level.

Selected Business Unit Profile by Country

Indicators	Country		
	Total	U.S.	Japan
Annual sales revenue[†]			
Mean	7,397.71	1,584.30	1,266.92
Standard deviation	48,471.56	6,177.42	3,838.27
Number	492	163	122
Pre-tax return on assets (% of assets)			
Mean	15.98	16.67	10.94
Standard deviation	19.96	15.05	12.74
Number	350	119	79
Net pre-tax profit ratio (% of sales)			
Mean	10.48	9.79	14.47
Standard deviation	32.01	9.55	63.27
Number	387	122	92
R&D expenses (% of sales)			
Mean	4.31	4.40	3.73
Standard deviation	3.97	4.56	2.98
Number	401	146	89
Growth rate in unit sales (%)			
Mean	22.17	7.60	53.52
Standard deviation	48.97	10.62	81.26
Number	470	156	118
Market share of primary product			
Mean	32.04	38.66	27.77
Standard deviation	21.74	22.32	17.70
Number	435	145	108
Capacity utilization (%)			
Mean	83.57	72.29	97.72
Standard deviation	22.41	18.30	19.41
Number	462	155	113
Number of plants in the business unit			
Mean	8.01	10.53	3.42
Standard deviation	17.93	16.25	4.40
Number	512	182	120
Number of employees			
Mean	6,041.02	8,747.12	2,089.38
Standard deviation	26,088.19	32,636.78	5,732.25
Number	528	175	138

| | Country | | | |
| | Europe | | | |
Total Europe	Latin	Germanic	Anglo-Saxon	Nordic
15,588.74*	36,758.47	3,904.09	399.05	1,259.48
73,789.70	114,779.66	12,893.34	639.31	1,696.24
207	81	47	39	40
18.05*	18.00	15.31	25.00	13.84
25.34	28.45	15.22	32.07	19.72
152	53	35	33	31
8.84	8.67	6.36	9.69	11.47
10.26	7.93	8.14	10.42	14.91
173	60	44	35	34
4.54	4.03	5.90	4.34	3.93
3.88	3.56	3.94	4.43	3.45
166	60	41	35	30
81.26*	22.04	10.98	10.65	9.73
29.99	45.58	11.90	10.59	11.81
196	75	44	40	37
29.30*	25.22	32.31	27.27	35.17
21.50	19.28	20.45	22.20	24.80
182	65	45	37	35
84.34*	86.30	87.00	79.73	82.41
22.23	28.21	13.19	19.05	21.41
194	70	46	41	37
8.44*	4.91	14.90	8.95	7.13
23.00	7.36	40.50	21.97	12.28
210	79	48	43	40
6,374.78	4,687.00	15,838.40	3,192.47	2,112.50
27,798.75	17,875.61	52,937.53	5,239.70	4,670.48
215	82	48	43	42

Selected Business Unit Profile by Country (*continued*)

		Country	
Indicators	*Total*	*U.S.*	*Japan*
Manufacturing direct labor employees			
Mean	2,282.46	2,469.11	1,169.26
Standard deviation	9,942.87	6,230.05	2,734.06
Number	476	156	125
Manufacturing indirect labor employees			
Mean	984.18	1,378.47	1,006.58
Standard deviation	2,694.76	3,862.31	3,418.92
Number	473	154	125

	Country			
	Europe			
Total Europe	Latin	Germanic	Anglo-Saxon	Nordic
2,846.73	985.35	8,488.42	1,501.72	1,129.58
14,322.03	1,436.83	29,102.03	2,651.17	2,617.60
195	75	45	39	36
656.75	573.55	743.83	892.88	471.27
1,388.66	1,037.54	1,855.54	1,678.04	1,016.43
194	75	42	40	37

[†]In U.S. dollars.

[*]The ANOVA F-test calculated from the comparison between the United States, Japan, and Europe is significant at the .05 alpha level.

(*concluded*)

Importance of Competitive Abilities by Country

Importance of Competitive Abilities	Country		
	Total	U.S.	Japan
Ability to profit in price-competitive market			
Mean	5.58	5.73	5.49
Standard deviation	1.29	1.28	1.27
Number	545	182	142
Ability to make rapid changes in design			
Mean	5.22	5.17	5.82
Standard deviation	1.42	1.37	1.11
Number	540	182	142
Ability to introduce new products quickly			
Mean	5.39	5.48	5.82
Standard deviation	1.35	1.19	1.11
Number	401	181	142
Ability to make rapid volume changes			
Mean	5.10	5.05	5.37
Standard deviation	1.41	1.40	1.19
Number	543	181	142
Ability to make rapid product mix changes			
Mean	5.10	5.37	5.04
Standard deviation	1.34	1.17	1.34
Number	536	182	135
Ability to offer a broad product line			
Mean	5.18	5.36	5.15
Standard deviation	1.40	1.33	1.33
Number	544	180	142
Ability to offer consistently low defect rates			
Mean	6.27	6.51	5.83
Standard deviation	1.00	0.73	1.16
Number	545	182	141
Ability to provide high-performance products or product amenities			
Mean	5.73	6.00	5.45
Standard deviation	1.20	1.02	1.22
Number	541	182	141
Ability to provide reliable/durable products			
Mean	6.15	6.31	6.17
Standard deviation	1.07	0.93	0.95
Number	535	180	141

	Country			
	Europe			
Total Europe	Latin	Germanic	Anglo-Saxon	Nordic
5.51	5.34	5.84	5.27	5.70
1.30	1.44	1.09	1.34	1.10
221	85	49	44	43
4.88*	5.08	5.27	4.33	4.55
1.52	1.48	1.43	1.60	1.45
216	83	49	42	42
5.31	5.31	5.75	5.05	5.05
1.48	1.61	1.29	1.54	1.23
220	84	51	43	42
4.96*	5.06	5.14	4.68	4.86
1.53	1.55	1.64	1.49	1.42
220	83	51	44	42
4.92*	5.02	4.86	5.07	4.64
1.44	1.43	1.66	1.42	1.14
219	83	51	43	42
5.05	4.95	5.43	4.68	5.14
1.49	1.52	1.39	1.54	1.42
222	85	51	44	42
6.36*	6.38	6.37	6.34	6.31
1.00	0.94	1.22	1.10	0.72
222	85	51	44	42
5.68*	5.70	5.82	5.93	5.24
1.28	1.29	1.42	1.11	1.14
218	83	51	42	42
6.00*	6.09	6.12	5.98	5.69
1.24	1.14	1.27	1.49	1.07
214	80	50	42	42

Importance of Competitive Abilities by Country (*continued*)

Importance of Competitive Abilities	Country		
	Total	*U.S.*	*Japan*
Ability to provide fast deliveries			
Mean	5.65	5.69	5.63
Standard deviation	1.11	1.05	1.17
Number	546	182	142
Ability to make dependable delivery promises			
Mean	6.18	6.34	5.92
Standard deviation	0.97	0.75	1.10
Number	544	182	141
Ability to provide effective after-sales service			
Mean	5.38	5.47	5.45
Standard deviation	1.42	1.34	1.15
Number	535	181	141
Ability to provide product support effectively			
Mean	5.45	5.65	5.38
Standard deviation	1.23	1.20	1.09
Number	534	182	141
Ability to make products easily available (broad distribution)			
Mean	5.16	5.54	5.04
Standard deviation	1.44	1.29	1.31
Number	535	181	138
Ability to customize products and services to customer needs			
Mean	5.62	5.47	5.88
Standard deviation	1.25	1.21	1.03
Number	538	182	139

| | Country | | | |
| | | Europe | | |
Total Europe	Latin	Germanic	Anglo-Saxon	Nordic
5.64	5.73	5.51	5.61	5.64
1.13	1.06	1.14	1.40	0.96
222	85	51	44	42
6.21*	6.06	6.33	6.45	6.10
1.01	1.20	0.74	1.00	0.82
221	84	51	44	42
5.25	5.27	5.79	5.02	4.85
1.62	1.60	1.50	1.49	1.80
213	82	47	43	41
5.33*	5.11	5.78	5.51	5.05
1.33	1.47	1.13	1.20	1.26
211	81	46	43	41
4.93*	4.63	5.48	4.95	4.85
1.58	1.83	1.38	1.33	1.35
216	84	48	44	40
5.57*	5.60	6.16	4.93	5.46
1.39	1.42	1.03	1.61	1.14
217	83	49	44	41

*The ANOVA F-test calculated from the comparison between the United States, Japan, and Europe is significant at the .05 alpha level.

(*concluded*)

Strength in Competitive Abilities by Country

Competitive Abilities	Country		
	Total	U.S.	Japan
Ability to profit in price-competitive market			
Mean	4.36	4.55	4.11
Standard deviation	1.34	1.39	1.33
Number	543	182	142
Ability to make rapid changes in design			
Mean	4.43	4.46	4.33
Standard deviation	1.27	1.27	1.20
Number	541	182	141
Ability to introduce new products quickly			
Mean	4.35	4.42	4.33
Standard deviation	1.29	1.27	1.20
Number	401	181	0
Ability to make rapid volume changes			
Mean	4.69	4.88	4.62
Standard deviation	1.23	1.28	1.10
Number	540	182	142
Ability to make rapid product mix changes			
Mean	4.80	5.00	4.33
Standard deviation	1.20	1.12	1.20
Number	399	181	0
Ability to offer a broad product line			
Mean	4.84	5.16	4.44
Standard deviation	1.45	1.37	1.28
Number	540	180	140
Ability to offer consistently low defect rates			
Mean	5.25	5.51	4.91
Standard deviation	1.17	1.11	1.14
Number	544	182	141
Ability to provide high-performance products or product amenities			
Mean	5.12	5.49	4.44
Standard deviation	1.22	1.16	1.16
Number	540	182	140
Ability to provide reliable/durable products			
Mean	5.44	5.80	5.06
Standard deviation	1.13	0.94	1.09
Number	537	181	140

Note: The Japanese tend to underrate their competitive abilities by .3 to .5 points compared to the Americans and Europeans. Benchmarkers may wish to adjust for this difference.

| | Country | | | |
| | Europe | | | |
Total Europe	Latin	Germanic	Anglo-Saxon	Nordic
4.37*	4.35	4.39	4.40	4.37
1.29	1.38	1.15	1.20	1.38
219	84	49	43	43
4.48	4.48	4.61	4.21	4.60
1.31	1.34	1.37	1.25	1.27
218	84	49	43	42
4.30	4.32	4.43	4.12	4.29
1.31	1.37	1.45	1.14	1.20
220	84	51	43	42
4.57*	4.62	4.65	4.27	4.68
1.25	1.39	1.05	1.09	1.31
216	82	49	44	41
4.64*	4.73	4.67	4.65	4.43
1.24	1.23	1.18	1.34	1.27
218	82	51	43	42
4.83*	4.49	5.20	4.95	4.93
1.55	1.60	1.44	1.51	1.55
220	83	51	44	42
5.26*	5.35	5.40	4.98	5.19
1.20	1.23	1.16	1.19	1.15
221	85	50	44	42
5.24*	5.17	5.37	5.43	5.05
1.14	1.31	1.04	0.86	1.10
218	83	51	42	42
5.40*	5.30	5.75	5.43	5.14
1.22	1.34	1.11	1.11	1.14
216	81	51	42	42

Strength in Competitive Abilities by Country (*continued*)

Competitive Abilities	*Total*	*U.S.*	*Japan*
Ability to provide fast deliveries			
Mean	4.82	5.18	4.46
Standard deviation	1.22	1.10	1.13
Number	545	182	142
Ability to make dependable delivery promises			
Mean	5.03	5.39	4.79
Standard deviation	1.31	1.24	1.27
Number	544	181	142
Ability to provide effective after-sales service			
Mean	5.01	5.43	4.62
Standard deviation	1.26	1.14	1.16
Number	533	181	142
Ability to provide product support effectively			
Mean	4.98	5.42	4.56
Standard deviation	1.19	1.06	1.11
Number	533	182	142
Ability to make products easily available (broad distribution)			
Mean	4.89	5.38	4.51
Standard deviation	1.31	1.30	1.11
Number	536	181	140
Ability to customize products and services to customer needs			
Mean	4.91	5.13	4.71
Standard deviation	1.23	1.15	1.13
Number	535	182	139

Total Europe	Country			
	Europe			
	Latin	Germanic	Anglo-Saxon	Nordic
4.75*	4.91	4.59	4.65	4.74
1.28	1.23	1.15	1.53	1.25
221	85	51	43	42
4.90*	4.92	4.94	4.55	5.19
1.34	1.30	1.33	1.47	1.23
221	84	51	44	42
4.90*	4.83	5.13	5.21	4.49
1.32	1.32	1.51	1.10	1.23
210	80	46	43	41
4.88*	4.63	5.09	5.20	4.78
1.23	1.21	1.18	1.25	1.21
209	79	45	44	41
4.72*	4.65	4.89	4.89	4.46
1.31	1.31	1.42	1.20	1.31
215	83	47	44	41
4.86*	4.89	5.11	4.64	4.76
1.33	1.42	1.06	1.45	1.28
214	82	46	44	42

*The ANOVA F-test calculated from the comparison between the United States, Japan, and Europe is significant at the .05 alpha level.

(*concluded*)

Manufacturing Cost Structure by Country

Manufacturing Cost Structure	Country		
	Total	U.S.	Japan
Manufacturing costs as percent of sales, 1989			
Mean	60.13	58.67	75.77
Standard deviation	21.88	20.54	13.66
Number	460	154	116
Manufacturing costs as percent of sales, 1992			
Mean	57.17	56.02	73.72
Standard deviation	21.37	18.94	12.58
Number	416	140	92
Materials cost (% of total mfg. cost), 1989			
Mean	55.60	56.58	N/A[†]
Standard deviation	19.36	16.78	N/A
Number	351	156	N/A
Materials cost (% of total mfg. cost), 1992			
Mean	57.84	58.35	N/A
Standard deviation	19.55	17.34	N/A
Number	325	139	N/A
Direct labor cost (% of total mfg. cost), 1989			
Mean	15.60	12.47	N/A
Standard deviation	11.35	8.48	N/A
Number	350	154	N/A
Direct labor cost (% of total mfg. cost), 1992			
Mean	14.62	11.72	N/A
Standard deviation	11.02	8.46	N/A
Number	323	137	N/A
Energy cost (% of total mfg. cost), 1989			
Mean	4.77	4.72	N/A
Standard deviation	6.12	4.84	N/A
Number	313	122	N/A
Energy cost (% of total mfg. cost), 1992			
Mean	4.84	4.88	N/A
Standard deviation	6.40	5.29	N/A
Number	297	112	N/A
Manufacturing overhead cost (% of total mfg. cost), 1989			
Mean	24.12	27.51	N/A
Standard deviation	14.19	13.51	N/A
Number	348	156	N/A

	Country			
	Europe			
Total Europe	Latin	Germanic	Anglo-Saxon	Nordic
51.76*	54.94	45.89	48.95	55.78
22.04	20.15	21.45	22.03	24.96
190	70	45	38	37
49.78*	52.39	43.52	47.76	54.76
22.15	21.21	21.09	21.21	25.07
184	69	44	37	34
54.82	51.45	53.11	59.68	58.82
21.21	23.74	17.84	19.01	21.31
195	74	47	40	34
57.46	54.97	54.42	62.41	61.35
21.10	23.17	18.24	18.47	22.27
186	71	45	39	31
18.05*	20.92	18.19	14.33	16.06
12.66	14.91	12.80	8.61	9.82
196	74	47	40	35
16.76*	18.96	17.30	13.18	15.50
12.17	14.23	12.08	8.71	10.05
186	71	44	39	32
4.81	4.46	3.98	5.43	5.91
6.83	6.65	4.44	5.87	10.21
191	71	46	40	34
4.81	4.27	4.02	5.74	5.91
7.00	6.77	3.74	6.25	10.85
185	70	44	39	32
21.36*	21.45	24.70	19.70	18.55
14.17	14.72	14.29	12.44	14.42
192	73	46	40	33

Manufacturing Cost Structure by Country (*continued*)

Manufacturing Cost Structure	Country		
	Total	U.S.	Japan
Manufacturing overhead cost (% of total mfg. cost), 1992			
Mean	23.09	26.16	N/A
Standard deviation	14.43	14.06	N/A
Number	322	139	N/A

	Country			
	Europe			
Total Europe	Latin	Germanic	Anglo-Saxon	Nordic
20.76*	21.39	23.07	18.77	18.50
14.31	15.42	12.85	12.06	16.24
183	70	44	39	30

*The ANOVA F-test calculated from the comparison between the United States, Japan, and Europe is significant at the .05 alpha level.
†N/A = not reported.

(*concluded*)

Manufacturing Performance Improvement Index by Country

	Country		
Performance Indicators	Total	U.S.	Japan
Overall quality as perceived by customers			
Mean	113.06	114.95	110.43
Standard deviation	16.36	17.73	14.34
Number	523	179	128
Average unit production costs for typical product			
Mean	106.65	105.88	105.42
Standard deviation	16.91	17.23	12.21
Number	524	179	132
Inventory turnover			
Mean	117.41	121.16	116.61
Standard deviation	43.95	60.97	41.02
Number	523	179	133
Speed of new product development and/or design change			
Mean	107.17	105.59	108.66
Standard deviation	17.14	16.18	15.03
Number	517	179	129
On-time delivery			
Mean	112.33	112.46	108.57
Standard deviation	30.22	20.05	24.67
Number	523	179	132
Equipment changeover (or set-up) time			
Mean	111.63	111.71	112.55
Standard deviation	20.14	20.31	16.36
Number	507	175	129
Market share			
Mean	107.41	106.19	103.52
Standard deviation	18.14	20.45	10.10
Number	516	179	127
Profitability			
Mean	121.21	115.39	118.37
Standard deviation	57.37	51.53	50.04
Number	496	175	122
Customer service			
Mean	111.94	111.83	107.94
Standard deviation	16.62	18.22	11.45
Number	507	179	124

	Country			
	Europe			
Total Europe	Latin	Germanic	Anglo-Saxon	Nordic
113.06	114.70	106.57	120.61	110.24
16.15	13.40	8.54	27.25	8.64
216	83	51	41	41
108.05	110.55	106.39	108.63	104.46
18.98	24.87	10.64	19.26	11.00
213	82	49	41	41
114.75	119.62	108.12	120.00	107.92
23.93	29.09	14.49	28.24	10.12
211	81	51	40	39
107.61	108.80	104.61	105.88	110.65
19.02	25.70	7.60	16.94	14.37
209	80	49	40	40
114.55	122.22	103.88	117.49	109.35
39.09	58.16	13.89	24.28	13.36
212	81	50	41	40
110.98	113.25	110.04	108.05	110.79
22.14	27.10	14.84	26.34	12.78
203	75	49	40	39
110.80*	110.34	109.43	113.18	111.15
19.26	11.52	19.82	23.60	25.77
210	80	51	39	40
128.06	129.44	114.96	123.08	145.08
65.43	46.25	35.28	45.30	117.25
199	75	45	39	40
114.48*	117.72	108.67	118.03	111.13
17.38	19.06	11.29	22.22	11.05
204	79	46	40	39

Manufacturing Performance Improvement Index by Country (*continued*)

Performance Indicators	Country		
	Total	U.S.	Japan
Manufacturing lead time			
Mean	110.25	108.29	108.16
Standard deviation	20.55	24.99	12.98
Number	521	179	131
Procurement lead time			
Mean	106.05	104.63	105.50
Standard deviation	16.54	20.11	11.30
Number	517	179	129
Delivery lead time			
Mean	107.94	104.85	108.66
Standard deviation	17.88	19.66	13.13
Number	519	179	129
Variety of products producible by manufacturing			
Mean	112.73	111.28	115.42
Standard deviation	22.21	18.89	21.23
Number	524	179	130

	Country			
	Europe			
Total Europe	Latin	Germanic	Anglo-Saxon	Nordic
113.20*	115.57	109.39	117.44	108.56
19.93	19.99	13.82	29.56	10.46
211	82	49	41	39
107.61	110.05	102.45	110.00	106.59
15.81	16.92	14.57	18.78	8.86
209	80	49	41	39
110.13*	111.69	102.59	117.07	109.08
18.51	18.63	13.15	24.80	12.72
211	81	49	41	40
112.32	115.05	109.55	109.75	112.73
25.13	30.39	16.49	23.51	23.96
215	83	51	40	41

*The ANOVA F-test calculated from the comparison between the United States, Japan, and Europe is significant at the .05 alpha level.

(*concluded*)

Manufacturing Objectives by Country

Manufacturing Objectives	Country		
	Total	U.S.	Japan
Improve conformance quality (reduce defects)			
Mean	5.95	6.04	6.04
Standard deviation	1.09	0.98	1.06
Number	547	182	143
Reduce unit costs			
Mean	5.84	5.71	6.24
Standard deviation	1.10	1.11	0.90
Number	548	182	143
Improve safety record			
Mean	4.69	4.38	5.39
Standard deviation	1.59	1.60	1.34
Number	543	182	141
Reduce manufacturing lead time			
Mean	5.30	5.15	5.81
Standard deviation	1.35	1.35	0.97
Number	545	182	143
Increase capacity			
Mean	4.47	3.80	5.48
Standard deviation	1.68	1.68	1.15
Number	546	182	143
Reduce procurement lead time			
Mean	4.99	4.91	5.46
Standard deviation	1.36	1.32	1.11
Number	542	182	142
Reduce new product development cycle			
Mean	5.45	5.55	5.70
Standard deviation	1.35	1.24	1.32
Number	544	182	141
Reduce materials costs			
Mean	5.37	5.48	5.50
Standard deviation	1.23	1.16	1.11
Number	543	182	141
Reduce overhead costs			
Mean	5.52	5.59	5.37
Standard deviation	1.19	1.14	1.16
Number	544	182	141
Improve direct labor productivity			
Mean	5.40	4.96	5.87
Standard deviation	1.23	1.33	0.96
Number	544	182	142

| | Country | | | |
| | | Europe | | |
Total Europe	Latin	Germanic	Anglo-Saxon	Nordic
5.83	5.75	6.14	5.82	5.62
1.10	1.09	1.02	1.19	1.10
222	85	51	44	42
5.70*	5.60	5.77	6.05	5.45
1.15	1.18	1.10	1.08	1.17
223	85	52	44	42
4.49*	4.61	4.59	4.66	3.95
1.60	1.55	1.53	1.80	1.50
220	83	51	44	42
5.08*	5.18	5.42	4.98	4.56
1.48	1.28	1.43	1.73	1.55
221	84	52	44	41
4.37*	4.43	4.46	4.07	4.48
1.66	1.59	1.42	1.85	1.89
222	84	52	44	42
4.76*	4.86	5.02	4.47	4.56
1.46	1.34	1.30	1.67	1.60
220	85	51	43	41
5.20*	5.08	5.67	5.05	5.05
1.43	1.54	1.11	1.64	1.20
221	85	51	44	41
5.19*	5.06	5.27	5.18	5.38
1.33	1.41	1.30	1.30	1.27
220	85	51	44	40
5.56	5.46	5.78	5.93	5.10
1.25	1.29	1.10	0.90	1.49
222	85	51	44	42
5.45*	5.32	5.66	5.43	5.46
1.19	1.23	1.17	1.25	1.07
220	85	50	44	41

Manufacturing Objectives by Country (*continued*)

Manufacturing Objectives	Total	U.S.	Japan
		Country	
Increase throughput			
Mean	5.44	5.31	5.76
Standard deviation	1.20	1.26	0.98
Number	543	182	141
Reduce number of vendors			
Mean	4.05	4.44	3.84
Standard deviation	1.46	1.39	1.38
Number	542	182	141
Improve vendor quality			
Mean	5.47	5.77	5.27
Standard deviation	1.18	1.09	1.07
Number	543	182	141
Reduce inventories			
Mean	5.19	5.31	5.34
Standard deviation	1.36	1.25	1.16
Number	546	182	143
Increase delivery reliability			
Mean	5.48	5.31	5.59
Standard deviation	1.28	1.25	1.17
Number	547	182	142
Increase delivery speed			
Mean	5.14	4.92	5.57
Standard deviation	1.28	1.34	1.02
Number	548	182	143
Improve ability to make rapid product mix changes			
Mean	4.92	4.68	5.42
Standard deviation	1.33	1.34	0.98
Number	544	182	140
Improve ability to make rapid volume changes			
Mean	4.70	4.45	5.16
Standard deviation	1.33	1.37	1.08
Number	545	182	142
Reduce break-even points			
Mean	5.09	4.93	5.84
Standard deviation	1.33	1.25	1.05
Number	540	182	142

	Country			
	Europe			
Total Europe	Latin	Germanic	Anglo-Saxon	Nordic
5.34*	5.41	5.42	5.25	5.19
1.26	1.25	1.23	1.35	1.23
220	82	52	44	42
3.86*	3.95	3.81	3.98	3.60
1.52	1.47	1.57	1.52	1.57
220	84	52	44	40
5.35*	5.58	5.38	5.32	4.83
1.27	1.02	1.32	1.49	1.34
220	84	52	44	40
5.01*	5.15	5.13	4.91	4.66
1.53	1.40	1.53	1.72	1.58
221	84	52	44	41
5.54	5.40	5.88	5.77	5.17
1.36	1.36	1.11	1.44	1.45
223	85	52	44	42
5.05*	5.06	5.46	4.95	4.62
1.31	1.28	1.16	1.49	1.25
223	85	52	44	42
4.79*	4.74	5.20	4.91	4.29
1.44	1.33	1.40	1.44	1.60
222	85	51	44	42
4.60*	4.74	5.00	4.25	4.21
1.37	1.33	1.13	1.43	1.49
221	84	51	44	42
4.72*	4.78	5.06	4.43	4.51
1.37	1.48	1.17	1.44	1.25
218	83	50	44	41

Manufacturing Objectives by Country (*continued*)

Manufacturing Objectives	*Total*	*U.S.*	*Japan*
		Country	
Raise employee morale			
Mean	5.15	5.07	5.42
Standard deviation	1.13	1.09	1.08
Number	548	182	143
Maximize cash flow			
Mean	5.11	5.09	5.09
Standard deviation	1.29	1.45	0.99
Number	535	182	134
Increase environmental safety and protection			
Mean	4.88	4.68	5.21
Standard deviation	1.46	1.49	1.31
Number	546	182	141
Reduce capacity			
Mean	2.26	2.42	2.09
Standard deviation	1.45	1.59	1.22
Number	544	182	141
Increase product or materials standardization			
Mean	4.68	4.52	4.96
Standard deviation	1.38	1.34	1.29
Number	544	182	141
Improve labor relations			
Mean	4.48	4.18	4.94
Standard deviation	1.37	1.42	1.09
Number	545	182	141
Improve white-collar productivity			
Mean	1.16	4.96	5.24
Standard deviation	5.14	1.13	1.15
Number	543	182	141
Increase range of products produced by existing facilities			
Mean	4.10	4.05	4.45
Standard deviation	1.53	1.57	1.22
Number	545	182	141
Meet financial shipping goals			
Mean	4.61	4.84	4.42
Standard deviation	1.43	1.39	1.32
Number	528	178	140

	Country			
		Europe		
Total Europe	Latin	Germanic	Anglo-Saxon	Nordic
5.04*	5.05	5.13	5.00	4.93
1.18	1.16	1.07	1.22	1.33
223	85	52	44	42
5.13	5.10	5.33	4.89	5.21
1.33	1.45	1.01	1.40	1.32
219	82	51	44	42
4.84*	4.58	5.08	5.11	4.79
1.50	1.51	1.44	1.51	1.47
223	85	52	44	42
2.23	2.00	2.73	2.20	2.10
1.45	1.24	1.56	1.64	1.41
221	84	51	44	42
4.64*	4.64	5.17	4.20	4.43
1.45	1.32	1.40	1.34	1.68
222	84	52	44	42
4.44*	4.58	4.62	4.00	4.41
1.42	1.37	1.24	1.68	1.36
222	85	52	44	41
5.23*	5.19	5.56	5.02	5.13
1.17	1.27	0.87	1.19	1.24
220	84	52	44	40
3.91*	3.99	4.27	3.61	3.60
1.64	1.56	1.63	1.60	1.75
222	84	52	44	42
4.55*	4.81	4.81	4.20	4.10
1.51	1.54	1.33	1.68	1.31
210	80	47	44	39

Manufacturing Objectives by Country (*continued*)

Manufacturing Objectives	Country		
	Total	U.S.	Japan
Improve pre-sales service and technical support			
Mean	4.54	4.16	4.92
Standard deviation	1.33	1.27	1.10
Number	530	182	136
Improve after-sales service			
Mean	4.68	4.40	4.96
Standard deviation	1.35	1.36	1.03
Number	533	182	137
Change culture of manufacturing organization			
Mean	5.13	5.30	4.99
Standard deviation	1.44	1.55	1.09
Number	543	182	140
Improve interfunctional communication			
Mean	5.43	5.52	5.20
Standard deviation	1.13	1.20	1.05
Number	545	182	141
Improve communication with external partners			
Mean	5.10	5.13	5.09
Standard deviation	1.27	1.39	1.05
Number	545	182	141
Reduce set-up/changeover times			
Mean	5.01	4.78	5.52
Standard deviation	1.42	1.38	1.05
Number	543	182	141

	Country			
	Europe			
Total Europe	Latin	Germanic	Anglo-Saxon	Nordic
4.62*	4.79	4.94	4.21	4.33
1.43	1.38	1.37	1.37	1.57
214	82	47	43	42
4.74*	4.79	5.21	4.35	4.50
1.48	1.49	1.30	1.33	1.70
215	82	48	43	42
5.08	5.21	4.71	5.43	4.90
1.54	1.39	1.64	1.65	1.51
222	84	52	44	42
5.49*	5.67	5.46	5.55	5.12
1.11	1.12	1.00	1.19	1.06
223	85	52	44	42
5.08	5.20	5.10	4.93	4.98
1.30	1.19	1.03	1.72	1.33
223	85	52	44	42
4.88*	4.77	5.38	4.77	4.56
1.57	1.48	1.59	1.58	1.63
220	83	52	44	41

*The ANOVA F-test calculated from the comparison between the United States, Japan, and Europe is significant at the .05 alpha level.

(concluded)

Past Pay-offs from Programs and Activities by Country

Programs / Activities	Total	U.S.	Japan
Giving workers a broad range of tasks and/or more responsibility			
Mean	4.24	4.35	4.35
Standard deviation	1.39	1.43	1.24
Number	440	136	132
Activity-based costing			
Mean	3.67	3.81	3.65
Standard deviation	1.54	1.50	1.48
Number	297	47	120
Manufacturing reorganization			
Mean	4.57	4.92	4.26
Standard deviation	1.41	1.34	1.37
Number	410	107	124
Worker training			
Mean	4.53	4.69	4.39
Standard deviation	1.28	1.33	1.19
Number	438	129	126
Management training			
Mean	4.60	5.57	5.58
Standard deviation	1.24	1.26	1.18
Number	421	112	126
Supervisor training			
Mean	4.61	4.72	4.43
Standard deviation	1.22	1.22	1.18
Number	422	123	124
Computer-aided manufacturing (CAM)			
Mean	4.13	4.55	4.09
Standard deviation	1.65	1.48	1.64
Number	359	76	119
Computer-aided design (CAD)			
Mean	4.47	4.79	4.88
Standard deviation	1.66	1.38	1.54
Number	405	119	130
Value analysis/product design			
Mean	4.29	4.68	4.69
Standard deviation	1.56	1.58	1.37
Number	330	65	127
Interfunctional work teams			
Mean	4.47	4.90	4.33
Standard deviation	1.33	1.34	1.08
Number	382	108	120

| | Country | | | |
| | Europe | | | |
Total Europe	Latin	Germanic	Anglo-Saxon	Nordic
4.06	4.16	3.71	3.92	4.39
1.46	1.28	1.51	1.66	1.52
172	69	34	36	33
3.63	3.96	3.83	2.52	3.75
1.61	1.52	1.65	1.44	1.55
130	50	29	23	28
4.58*	4.81	4.39	4.49	4.41
1.44	1.19	1.60	1.50	1.62
179	68	38	39	34
4.52	4.69	4.19	4.71	4.37
1.30	1.21	1.35	1.25	1.41
183	68	36	38	41
4.62	4.63	4.85	4.38	4.62
1.26	1.20	1.08	1.40	1.41
183	67	40	37	39
4.66	4.65	4.86	4.66	4.51
1.23	1.37	1.09	1.02	1.30
175	65	35	38	37
3.96*	4.25	4.09	3.62	3.55
1.69	1.52	1.88	1.61	1.82
164	67	35	29	33
3.90*	3.59	4.63	3.74	3.84
1.79	1.77	1.57	1.77	1.95
156	63	35	27	31
3.75*	3.75	4.21	3.67	3.25
1.56	1.54	1.56	1.28	1.69
138	56	33	21	28
4.27*	4.44	4.26	4.32	3.86
1.43	1.26	1.41	1.66	1.51
154	61	31	34	28

Past Pay-offs from Programs and Activities by Country (*continued*)

Programs / Activities	Total	U.S.	Japan
		Country	
Quality function deployment			
Mean	4.63	4.61	4.81
Standard deviation	1.38	1.44	1.23
Number	384	90	120
Developing new processes for new products			
Mean	4.65	4.71	4.90
Standard deviation	1.42	1.28	1.25
Number	388	96	128
Developing new processes for old products			
Mean	4.59	4.66	5.08
Standard deviation	1.49	1.35	1.17
Number	370	83	129
Integrating information systems in manufacturing			
Mean	4.34	4.04	4.52
Standard deviation	1.39	1.45	1.27
Number	412	110	128
Integrating information systems across functions			
Mean	4.24	4.14	4.32
Standard deviation	1.42	1.39	1.33
Number	383	94	124
Reconditioning physical plants			
Mean	4.33	4.77	4.09
Standard deviation	1.48	1.44	1.27
Number	342	75	114
Just-in-time			
Mean	4.38	4.83	4.22
Standard deviation	1.67	1.59	1.63
Number	380	100	126
Robots			
Mean	3.63	3.33	4.32
Standard deviation	1.89	1.65	1.64
Number	293	42	126
Flexible manufacturing systems			
Mean	3.98	4.13	4.01
Standard deviation	1.56	1.49	1.29
Number	328	61	126

Total Europe	Country			
	Europe			
	Latin	Germanic	Anglo-Saxon	Nordic
4.52	4.61	4.74	4.38	4.19
1.43	1.46	1.31	1.48	1.47
174	67	42	34	31
4.43*	4.37	4.82	4.62	3.94
1.58	1.28	1.69	1.76	1.70
164	62	38	29	35
4.16*	4.15	4.52	3.90	4.09
1.66	1.66	1.79	1.51	1.68
158	62	31	31	34
4.41*	4.54	4.69	3.93	4.23
1.41	1.46	1.17	1.28	1.59
174	70	39	30	35
4.23	4.32	4.46	4.06	3.94
1.50	1.56	1.32	1.29	1.77
165	66	37	31	31
4.29*	4.69	3.85	4.26	3.97
1.59	1.47	1.62	1.87	1.40
153	62	33	27	31
4.21*	4.56	4.18	3.54	4.19
1.71	1.73	1.57	1.90	1.56
154	61	34	28	31
3.04*	3.25	3.28	2.50	2.73
1.98	1.94	2.07	2.04	1.91
125	52	29	18	26
3.90	3.95	4.28	3.48	3.62
1.80	1.70	1.73	2.02	1.88
144	58	36	21	29

Past Pay-offs from Programs and Activities by Country (*continued*)

	Country		
Programs / Activities	*Total*	*U.S.*	*Japan*
Design for manufacture			
Mean	4.03	4.46	4.25
Standard deviation	1.54	1.61	1.15
Number	316	80	118
Statistical quality control			
Mean	4.49	4.88	4.37
Standard deviation	1.45	1.54	1.13
Number	407	127	118
Closing and/or relocating plants			
Mean	3.24	4.62	2.70
Standard deviation	2.14	2.01	2.05
Number	297	61	115
Quality circles			
Mean	4.25	4.31	4.88
Standard deviation	1.68	1.65	1.25
Number	340	65	129
Investing in improved production-inventory control systems			
Mean	4.31	4.24	4.61
Standard deviation	1.54	1.63	1.31
Number	366	89	130
Hiring in new skills from outside			
Mean	3.89	4.29	3.79
Standard deviation	1.55	1.55	1.29
Number	333	68	118
Linking manufacturing strategy to business strategy			
Mean	4.62	4.85	4.43
Standard deviation	1.45	1.35	1.39
Number	422	133	120

	Country			
	Europe			
Total Europe	Latin	Germanic	Anglo-Saxon	Nordic
3.52*	3.53	3.68	3.43	3.41
1.70	1.64	1.99	1.67	1.60
118	43	25	23	27
4.25*	4.03	4.68	4.50	3.88
1.52	1.52	1.46	1.41	1.58
162	60	40	30	32
3.07*	2.82	3.10	3.16	3.36
2.14	2.03	1.92	2.36	2.06
121	45	29	19	28
3.67*	3.83	4.12	3.50	2.96
1.82	1.86	1.71	1.77	1.76
146	58	33	28	27
4.09*	4.00	4.13	4.33	4.00
1.65	1.61	1.75	1.59	1.71
147	58	31	27	31
3.78*	4.23	3.50	3.73	3.14
1.72	1.44	1.98	1.93	1.56
147	61	32	26	28
4.57	4.63	4.63	4.94	4.11
1.54	1.45	1.53	1.37	1.78
169	62	38	32	37

*The ANOVA F-test calculated from the comparison between the United States, Japan, and Europe is significant at the .05 alpha level.

(concluded)

Future Emphasis on Programs and Activities by Country

	Country		
Programs / Activities	*Total*	*U.S.*	*Japan*
Giving workers a broad range of tasks and/or more responsibilities			
Mean	5.07	5.46	4.98
Standard deviation	1.47	1.27	1.18
Number	505	173	129
Activity-based costing			
Mean	4.13	3.87	4.28
Standard deviation	1.68	1.73	1.43
Number	460	165	116
Manufacturing reorganization			
Mean	4.57	4.27	4.68
Standard deviation	1.57	1.68	1.17
Number	495	173	120
Worker training			
Mean	5.23	5.41	4.97
Standard deviation	1.16	1.15	1.11
Number	504	175	124
Management training			
Mean	5.18	5.12	5.29
Standard deviation	1.27	1.37	1.09
Number	506	174	125
Supervisor training			
Mean	5.23	5.23	5.09
Standard deviation	1.20	1.26	1.13
Number	499	174	122
Computer-aided manufacturing (CAM)			
Mean	4.54	4.16	5.07
Standard deviation	1.85	1.81	1.66
Number	483	171	119
Computer-aided design (CAD)			
Mean	4.72	4.63	5.37
Standard deviation	1.73	1.62	1.34
Number	500	175	129
Value analysis/product redesign			
Mean	4.53	4.27	5.26
Standard deviation	1.66	1.70	1.27
Number	469	164	123
Interfunctional work teams			
Mean	4.98	5.19	4.72
Standard deviation	1.50	1.61	1.25
Number	478	171	116

| | Country | | | |
| | Europe | | | |
Total Europe	Latin	Germanic	Anglo-Saxon	Nordic
4.80*	4.86	4.77	4.88	4.64
1.70	1.71	1.21	1.97	1.86
203	78	43	43	39
4.26	4.21	4.84	4.00	3.97
1.77	1.92	1.55	1.76	1.64
179	70	38	36	35
4.77	4.90	4.79	4.62	4.63
1.64	1.67	1.63	1.70	1.57
202	79	43	42	38
5.24	5.10	5.07	5.60	5.32
1.17	1.42	0.87	1.06	0.96
205	78	44	42	41
5.17	5.09	5.55	5.15	4.95
1.29	1.39	0.90	1.24	1.45
207	80	44	41	42
5.32	5.28	5.30	5.48	5.23
1.19	1.36	0.94	1.02	1.27
203	78	43	42	40
4.54*	4.54	4.88	4.24	4.43
1.92	2.12	1.80	1.72	1.82
193	76	43	37	37
4.37*	4.33	5.17	3.79	4.18
1.93	2.04	1.66	1.91	1.79
196	78	41	38	39
4.26*	4.35	4.83	3.69	3.97
1.72	1.86	1.39	1.71	1.58
182	75	40	35	32
4.96	4.90	5.26	5.25	4.44
1.51	1.59	1.20	1.35	1.68
191	77	38	40	36

Future Emphasis on Programs and Activities by Country (*continued*)

Programs / Activities	Total	U.S.	Japan
		Country	
Quality function deployment			
Mean	5.20	5.02	5.31
Standard deviation	1.44	1.47	1.19
Number	484	169	118
Developing new processes for new products			
Mean	5.16	4.98	5.71
Standard deviation	1.56	1.60	1.09
Number	500	171	126
Developing new processes for old products			
Mean	4.56	4.11	5.54
Standard deviation	1.70	1.66	1.10
Number	488	171	127
Integrating information systems in manufacturing			
Mean	5.31	5.00	5.77
Standard deviation	1.33	1.35	1.06
Number	506	175	127
Integrating information systems across functions			
Mean	5.18	4.99	5.50
Standard deviation	1.38	1.44	1.12
Number	494	175	123
Reconditioning physical plants			
Mean	4.25	3.68	4.83
Standard deviation	1.66	1.66	1.38
Number	467	164	115
Just-in-time			
Mean	4.88	4.91	4.85
Standard deviation	1.64	1.66	1.54
Number	483	167	123
Robots			
Mean	3.45	2.54	5.11
Standard deviation	2.01	1.50	1.43
Number	460	161	124
Flexible manufacturing systems			
Mean	4.45	4.03	5.02
Standard deviation	1.83	1.78	1.51
Number	473	165	122

	Country			
	Europe			
Total Europe	Latin	Germanic	Anglo-Saxon	Nordic
5.30	5.39	5.88	5.23	4.46
1.55	1.48	1.03	1.59	1.84
197	79	43	40	35
4.98*	4.82	5.43	5.00	4.71
1.70	1.73	1.43	1.91	1.68
203	76	47	42	38
4.31*	4.01	4.60	4.34	4.56
1.81	1.90	1.61	1.91	1.70
190	76	40	38	36
5.29*	5.40	5.54	5.12	4.95
1.38	1.57	1.11	1.33	1.30
204	77	46	41	40
5.14	5.26	5.40	5.25	4.47
1.44	1.60	1.14	1.24	1.48
196	77	43	40	36
4.40*	4.66	4.46	4.14	4.06
1.67	1.72	1.48	1.64	1.72
188	76	39	37	36
4.88	4.90	5.21	4.68	4.63
1.69	1.84	1.49	1.68	1.57
193	77	43	38	35
3.12*	3.35	3.66	2.26	2.83
2.06	2.13	2.02	1.77	2.01
175	71	38	31	35
4.45*	4.55	4.85	4.17	4.08
1.98	2.00	1.93	2.06	1.86
186	73	41	36	36

Future Emphasis on Programs and Activities by Country (*continued*)

Programs / Activities	Total	U.S.	Japan
		Country	
Design for manufacture			
Mean	4.73	4.95	5.28
Standard deviation	1.79	1.73	1.11
Number	453	165	120
Statistical quality control			
Mean	5.02	5.40	4.69
Standard deviation	1.50	1.42	1.16
Number	493	177	118
Closing and/or relocating plants			
Mean	2.93	3.05	3.01
Standard deviation	2.09	2.13	2.12
Number	445	159	116
Quality circles			
Mean	4.21	3.52	5.04
Standard deviation	1.91	1.93	1.21
Number	466	159	127
Investing in improved production-inventory control systems			
Mean	4.87	4.55	5.58
Standard deviation	1.60	1.51	1.10
Number	483	168	130
Hiring in new skills from outside			
Mean	4.06	3.78	4.66
Standard deviation	1.71	1.64	1.30
Number	475	166	119
Linking manufacturing strategy to business strategy			
Mean	5.58	5.60	5.54
Standard deviation	1.24	1.24	1.19
Number	499	177	121

	Country			
	Europe			
Total Europe	Latin	Germanic	Anglo-Saxon	Nordic
4.11*	4.18	4.45	3.74	4.03
2.06	2.14	1.84	2.17	2.04
168	65	33	35	35
4.87*	4.70	5.27	5.18	4.41
1.67	1.86	1.42	1.30	1.77
198	76	45	40	37
2.76	2.63	3.33	2.61	2.63
2.04	2.07	2.10	2.06	1.91
170	64	33	38	35
4.23*	4.63	4.92	3.81	3.17
2.07	1.85	1.81	2.27	2.09
180	72	36	37	35
4.66*	4.69	4.92	4.60	4.43
1.82	1.94	1.59	1.71	1.93
185	74	36	40	35
3.93*	4.64	3.69	3.21	3.50
1.90	1.78	2.02	1.88	1.56
190	75	42	39	34
5.59	5.47	5.62	6.02	5.33
1.28	1.35	1.31	0.87	1.42
201	78	42	42	39

*The ANOVA F-test calculated from the comparison between the United States, Japan, and Europe is significant at the .05 alpha level.

(concluded)

Internal Environment by Country

		Country	
Indicators	Total	U.S.	Japan
Top management only			
Mean	0.03	0.01	0.01
Standard deviation	0.18	0.07	0.08
Number	549	182	143
Top and some middle management			
Mean	0.26	0.24	0.10
Standard deviation	0.44	0.43	0.31
Number	549	182	143
Top and most middle management			
Mean	0.31	0.43	0.34
Standard deviation	0.46	0.50	0.48
Number	549	182	143
Every manager and supervisor			
Mean	0.26	0.20	0.36
Standard deviation	0.44	0.40	0.48
Number	549	182	143
Every manager, supervisor, and worker			
Mean	0.13	0.13	0.17
Standard deviation	0.34	0.34	0.38
Number	549	182	143

	Country			
	Europe			
Total Europe	Latin	Germanic	Anglo-Saxon	Nordic
0.07*	0.06	0.06	0.05	0.14
0.26	0.24	0.24	0.21	0.35
224	85	52	44	43
0.38*	0.36	0.40	0.36	0.37
0.49	0.48	0.50	0.49	0.49
224	85	52	44	43
0.20*	0.21	0.21	0.14	0.21
0.40	0.41	0.41	0.35	0.41
224	85	52	44	43
0.23*	0.24	0.29	0.23	0.16
0.42	0.43	0.46	0.42	0.37
224	85	52	44	43
0.11	0.12	0.04	0.20	0.07
0.31	0.32	0.19	0.41	0.26
224	85	52	44	43

*The ANOVA F-test calculated from the comparison between the United States, Japan, and Europe is significant at the .05 alpha level.

Past and Future Percentage of Domestic and Foreign Sales, Productions, and Purchases by Country

	Country		
Programs / Activities	Total	U.S.	Japan
Domestic sales, 1989			
Mean	83.65	85.41	82.91
Standard deviation	38.60	59.94	17.96
Number	458	160	116
Foreign sales, 1989			
Mean	19.27	21.39	17.55
Standard deviation	116.00	35.61	18.81
Number	458	160	116
Domestic sales, 1992			
Mean	83.34	82.42	88.86
Standard deviation	43.71	69.18	16.59
Number	445	155	118
Foreign sales, 1992			
Mean	19.66	25.90	10.97
Standard deviation	30.43	43.43	15.81
Number	444	155	117
Domestic production, 1989			
Mean	92.22	89.30	93.31
Standard deviation	14.37	16.12	12.61
Number	454	155	115
Foreign production, 1989			
Mean	8.08	10.70	7.57
Standard deviation	15.25	16.12	15.30
Number	453	155	115
Domestic production, 1992			
Mean	90.29	85.94	91.79
Standard deviation	16.35	19.21	14.29
Number	440	152	111
Foreign production, 1992			
Mean	9.90	14.06	9.02
Standard deviation	16.91	19.21	16.73
Number	439	152	111
Domestic purchases, 1989			
Mean	84.80	84.25	89.23
Standard deviation	20.45	18.18	18.17
Number	438	153	109

	Country			
	Europe			
Total Europe	Latin	Germanic	Anglo-Saxon	Nordic
82.59	86.61	80.00	79.11	80.72
19.93	17.46	19.79	22.78	21.01
182	69	41	36	36
18.51	15.39	19.12	20.83	21.44
21.15	20.92	19.61	22.73	21.76
182	69	41	36	36
80.39	85.11	78.56	76.48	77.34
20.77	17.48	20.50	24.31	22.30
172	65	39	33	35
19.95*	15.03	21.08	23.52	24.49
21.04	17.44	20.34	24.29	23.57
172	65	39	33	35
94.00*	96.50	87.67	95.11	95.46
13.50	9.44	19.29	12.74	10.72
184	68	42	35	39
6.17*	3.51	13.22	4.89	4.54
14.18	9.44	21.25	12.74	10.72
183	68	41	35	39
93.07*	95.78	88.39	92.94	93.61
14.05	9.99	18.03	15.71	12.80
177	65	41	33	38
6.86*	4.23	11.43	7.06	6.39
14.03	10.00	18.10	15.71	12.80
176	65	40	33	38
82.53*	85.22	79.15	79.12	84.48
23.16	20.93	25.77	26.10	21.20
176	63	39	34	40

Past and Future Percentage of Domestic and Foreign Sales, Productions, and Purchases by Country (*continued*)

Programs / Activities	Country		
	Total	*U.S.*	*Japan*
Foreign purchases, 1989			
Mean	15.38	15.75	11.61
Standard deviation	20.86	18.18	20.08
Number	437	153	109
Domestic purchases, 1992			
Mean	76.65	82.21	57.70
Standard deviation	27.49	17.88	38.45
Number	419	150	98
Foreign purchases, 1992			
Mean	27.87	17.79	54.04
Standard deviation	31.73	17.88	40.16
Number	449	150	128

	Country			
	Europe			
Total Europe	Latin	Germanic	Anglo-Saxon	Nordic
17.42	14.79	20.90	20.59	15.40
23.21	21.09	25.76	26.17	21.12
175	62	39	34	40
82.63*	84.81	78.82	80.97	84.21
21.55	21.52	22.39	23.24	19.45
171	62	38	32	39
17.11*	14.44	21.21	19.09	15.74
21.12	20.11	22.38	23.38	19.4689
171	62	38	32	39

*The ANOVA F-test calculated from the comparison between the United States, Japan, and Europe is significant at the .05 alpha level.

(*concluded*)

APPENDIX B

INDUSTRY DATA

Business Unit Growth Strategy by Industry

	Industry		
Indicators	Total	Consumer	Industrial Goods
Build market share			
Mean	0.70	0.78	0.81
Standard deviation	0.46	0.41	0.40
Number	539	88	57
Hold (defend) market share			
Mean	0.27	0.19	0.18
Standard deviation	0.44	0.40	0.38
Number	539	88	57
Harvest (maximize cash flow, sacrifice market share)			
Mean	0.03	0.01	0.02
Standard deviation	0.16	0.11	0.13
Number	539	88	57
Withdraw (prepare to exit the business)			
Mean	0.00	0.00	0.00
Standard deviation	0.06	0.00	0.00
Number	539	88	57

	Industry	
Basic	Machinery	Electronics
0.56	0.69	0.72*
0.50	0.46	0.45
126	133	135
0.41	0.26	0.24*
0.49	0.44	0.43
126	133	135
0.02	0.04	0.04
0.15	0.19	0.19
126	133	135
0.00	0.01	0.01
0.00	0.09	0.09
126	133	135

*The ANOVA F-test calculated from the comparison between industries is significant at the .05 alpha level.

Selected Business Unit Profit by Industry

	Industry	
Indicators	*Total*	*Consumer*
Annual sales revenue[†]		
Mean	7,484.31	7,178.63
Standard deviation	48,814.58	27,359.92
Number	485	76
Pre-tax return on assets (% of assets)		
Mean	16.00	18.17
Standard deviation	20.10	14.95
Number	345	58
Net pre-tax profit ratio (% of sales)		
Mean	10.45	9.81
Standard deviation	32.21	9.36
Number	382	57
R&D expenses (% of sales)		
Mean	4.29	2.98
Standard deviation	3.97	3.86
Number	395	62
Growth rate in unit sales (%)		
Mean	21.89	20.93
Standard deviation	49.10	47.16
Number	462	74
Market share of primary product		
Mean	32.11	35.48
Standard deviation	21.76	22.52
Number	430	69
Capacity utilization (%)		
Mean	83.52	79.16
Standard deviation	22.46	30.52
Number	456	75
Number of plants in the business unit		
Mean	8.08	9.20
Standard deviation	18.08	10.54
Number	503	84
Number of employees		
Mean	6,117.69	3,426.67
Standard deviation	26,325.64	5,298.98
Number	518	84
Manufacturing direct labor employees		
Mean	2,302.10	1,469.65
Standard deviation	10,030.70	2,500.96
Number	467	75

	Industry		
Industrial Goods	*Basic*	*Machinery*	*Electronics*
309.40	15,680.89	5,950.38	4,922.77
435.57	92,486.97	22,271.41	26,260.80
53	111	118	127
17.43	17.25	10.49	17.65
16.10	18.73	12.37	29.32
46	77	78	86
8.62	11.31	13.54	8.14
5.90	13.37	63.05	10.20
47	85	92	101
2.85	3.73	3.54	6.77*
2.54	3.12	2.49	5.00
48	84	93	108
9.36	22.73	20.79	28.10
11.36	76.68	31.88	40.61
50	107	113	118
34.49	28.01	33.14	31.54
23.57	22.02	21.22	20.52
47	97	110	107
74.49	85.51	88.11	84.26*
18.90	21.39	22.38	16.53
53	107	111	110
6.18	12.57	5.72	6.38*
8.08	30.41	10.11	15.61
57	115	125	122
1,566.77	2,707.53	11,537.96	7,674.83*
2,715.92	5,221.21	36,970.05	36,601.14
57	120	128	129
770.39	1,037.66	4,906.83	1,955.80*
1,522.56	1,856.34	18,232.53	6,713.43
51	104	120	117

Selected Business Unit Profit by Industry (*continued***)**

	Industry	
Indicators	*Total*	*Consumer*
Manufacturing indirect labor employees		
Mean	984.01	498.49
Standard deviation	2,978.19	881.78
Number	464	75

	Industry		
Industrial Goods	*Basic*	*Machinery*	*Electronics*
304.78	638.72	1,533.15	1,340.26[*]
554.46	1,351.94	4,165.45	3,844.93
50	104	117	118

[†]In U.S. dollars.

[*]The ANOVA F-test calculated from the comparison between industries is significant at the .05 alpha level.

(*concluded*)

Importance of Competitive Abilities by Industry

	Industry	
Importance of Competitive Abilities	*Total*	*Consumer*
Ability to profit in price competitive market		
Mean	5.58	5.39
Standard deviation	1.29	1.50
Number	535	87
Ability to make rapid changes in design		
Mean	5.22	4.74
Standard deviation	1.42	1.57
Number	530	85
Ability to introduce new products quickly		
Mean	5.39	5.46
Standard deviation	1.35	1.30
Number	401	79
Ability to make rapid volume changes		
Mean	5.09	5.07
Standard deviation	1.41	1.41
Number	533	88
Ability to make rapid product mix changes		
Mean	5.10	4.99
Standard deviation	1.33	1.26
Number	526	88
Ability to offer a broad product line		
Mean	5.18	5.13
Standard deviation	1.41	1.40
Number	534	88
Ability to offer consistently low defect rates		
Mean	6.28	6.19
Standard deviation	0.99	1.23
Number	535	88
Ability to provide high-performance products or product amenities		
Mean	5.73	5.54
Standard deviation	1.21	1.52
Number	531	85
Ability to provide reliable/durable products		
Mean	6.15	6.00
Standard deviation	1.07	1.17
Number	525	83
Ability to provide fast deliveries		
Mean	5.65	5.74
Standard deviation	1.12	1.21
Number	536	88

	Industry		
Industrial Goods	*Basic*	*Machinery*	*Electronics*
5.60	5.48	5.58	5.77
1.40	1.18	1.28	1.17
57	126	130	135
4.98	4.71	5.73	5.59*
1.41	1.63	1.06	1.15
57	124	130	134
5.12	4.87	5.60	5.81*
1.30	1.54	1.18	1.21
57	90	84	91
5.35	4.85	5.12	5.19
1.34	1.50	1.32	1.44
55	126	130	134
5.16	4.84	5.22	5.28
1.39	1.46	1.22	1.34
57	120	129	132
5.25	5.10	5.31	5.14
1.39	1.41	1.46	1.37
57	123	131	135
6.58	6.21	6.32	6.24
0.68	0.94	0.97	0.99
57	125	130	135
5.89	5.49	5.90	5.82*
0.99	1.25	1.04	1.14
57	123	131	135
6.19	5.81	6.35	6.35*
1.03	1.28	0.93	0.84
57	122	130	133
5.60	5.65	5.63	5.63
1.08	1.12	1.12	1.06
57	125	131	135

Importance of Competitive Abilities by Industry (*continued*)

	Industry	
Importance of Competitive Abilities	*Total*	*Consumer*
Ability to make dependable delivery promises		
Mean	6.19	6.20
Standard deviation	0.95	1.05
Number	534	87
Ability to provide effective after-sales service		
Mean	5.40	4.70
Standard deviation	1.42	1.68
Number	526	83
Ability to provide product support effectively		
Mean	5.46	4.98
Standard deviation	1.23	1.38
Number	525	83
Ability to make products easily available (broad distribution)		
Mean	5.17	5.46
Standard deviation	1.45	1.46
Number	526	87
Ability to customize products and services to customer needs		
Mean	5.61	5.17
Standard deviation	1.25	1.50
Number	529	87

	Industry		
Industrial Goods	Basic	Machinery	Electronics
6.35	6.15	6.20	6.16
0.79	1.04	0.83	0.97
57	124	131	135
4.84	5.15	5.95	5.76*
1.47	1.41	1.14	1.12
57	123	130	133
5.23	5.30	5.85	5.65*
1.26	1.29	1.11	1.04
56	123	130	133
5.33	5.06	5.08	5.08
1.54	1.35	1.49	1.45
55	124	129	131
5.32	5.56	5.89	5.80*
1.33	1.27	0.98	1.15
57	123	130	132

*The ANOVA F-test calculated from the comparison between industries is significant at the .05 alpha level.

(concluded)

Strength in Competitive Abilities by Industry

	Industry	
Competitive Abilities	*Total*	*Consumer*
Ability to profit in price-competitive market		
Mean	4.36	4.61
Standard deviation	1.35	1.40
Number	533	85
Ability to make rapid changes in design		
Mean	4.44	4.15
Standard deviation	1.27	1.37
Number	531	85
Ability to introduce new products quickly		
Mean	4.35	4.18
Standard deviation	1.29	1.35
Number	401	79
Ability to make rapid volume changes		
Mean	4.69	4.70
Standard deviation	1.23	1.27
Number	530	86
Ability to make rapid product mix changes		
Mean	4.80	4.84
Standard deviation	1.20	1.15
Number	399	79
Ability to offer a broad product line		
Mean	4.85	4.94
Standard deviation	1.45	1.66
Number	530	87
Ability to offer consistently low defect rates		
Mean	5.26	5.43
Standard deviation	1.17	1.22
Number	534	86
Ability to provide high-performance products or product amenities		
Mean	5.13	5.12
Standard deviation	1.23	1.20
Number	531	84
Ability to provide reliable/durable products		
Mean	5.45	5.41
Standard deviation	1.13	1.21
Number	527	83
Ability to provide fast deliveries		
Mean	4.83	4.92
Standard deviation	1.22	1.19
Number	535	87

Industry			
Industrial Goods	*Basic*	*Machinery*	*Electronics*
4.58	4.55	4.08	4.21*
1.46	1.30	1.25	1.35
57	125	131	135
4.82	4.41	4.43	4.50*
1.17	1.39	1.21	1.16
57	124	132	133
4.49	4.51	4.43	4.20
1.28	1.33	1.25	1.25
57	90	84	91
4.98	4.49	4.73	4.72
1.26	1.31	1.18	1.17
56	125	131	132
4.91	4.70	4.77	4.84
1.24	1.24	1.24	1.18
57	89	84	90
5.09	4.82	4.74	4.83
1.34	1.41	1.49	1.35
56	122	132	133
5.47	5.09	5.08	5.38*
1.24	1.13	1.07	1.21
57	125	131	135
5.47	4.85	5.15	5.21*
1.18	1.21	1.21	1.25
57	123	132	135
5.75	5.23	5.49	5.51*
0.99	1.20	1.04	1.13
57	122	131	134
5.19	4.93	4.65	4.70*
1.26	1.16	1.31	1.15
57	124	132	135

Strength in Competitive Abilities by Industry (*continued*)

	Industry	
Competitive Abilities	*Total*	*Consumer*
Ability to make dependable delivery promises		
Mean	5.05	5.06
Standard deviation	1.32	1.34
Number	534	86
Ability to provide effective after-sales service		
Mean	5.02	4.93
Standard deviation	1.26	1.30
Number	523	81
Ability to provide product support effectively		
Mean	4.99	4.91
Standard deviation	1.19	1.25
Number	523	80
Ability to make products easily available (broad distribution)		
Mean	4.90	5.29
Standard deviation	1.31	1.29
Number	526	85
Ability to customize products and services to customer needs		
Mean	4.92	4.65
Standard deviation	1.24	1.17
Number	525	85

	Industry		
Industrial Goods	Basic	Machinery	Electronics
5.49	5.06	5.02	4.87
1.10	1.26	1.38	1.34
57	124	132	135
5.23	4.79	5.07	5.17
1.21	1.31	1.30	1.16
57	122	131	132
5.19	4.95	5.02	4.97
1.39	1.11	1.18	1.14
57	122	131	133
5.24	4.73	4.80	4.75*
1.39	1.26	1.35	1.21
55	123	131	132
5.35	4.77	5.02	4.93*
1.16	1.35	1.23	1.16
57	121	130	132

*The ANOVA F-test calculated from the comparison between industries is significant at the .05 alpha level.

(*concluded*)

Manufacturing Cost Structure by Industry

Manufacturing Cost Structure	Industry	
	Total	Consumer
Manufacturing costs as percent of sales, 1989		
Mean	59.96	48.84
Standard deviation	21.88	22.46
Number	452	77
Manufacturing costs as percent of sales, 1992		
Mean	56.98	45.73
Standard deviation	21.35	22.45
Number	412	71
Materials cost (% of total mfg. cost), 1989		
Mean	55.60	57.01
Standard deviation	19.36	19.80
Number	351	67
Materials cost (% of total mfg. cost), 1992		
Mean	57.84	56.86
Standard deviation	19.55	19.74
Number	325	63
Direct labor cost (% of total mfg. cost), 1989		
Mean	15.60	16.99
Standard deviation	11.35	12.23
Number	350	67
Direct labor cost (% of total mfg. cost), 1992		
Mean	14.62	16.79
Standard deviation	11.02	11.97
Number	323	62
Energy cost (% of total mfg. cost), 1989		
Mean	4.77	3.87
Standard deviation	6.12	3.63
Number	313	63
Energy cost (% of total mfg. cost), 1992		
Mean	4.84	4.00
Standard deviation	6.40	3.48
Number	297	59
Manufacturing overhead cost (% of total mfg. cost), 1989		
Mean	24.12	22.59
Standard deviation	14.19	13.42
Number	348	66

	Industry		
Industrial Goods	*Basic*	*Machinery*	*Electronics*
59.84	62.21	68.13	57.42*
20.98	22.56	21.16	18.31
51	101	112	111
57.45	60.45	64.09	54.32*
20.33	22.31	20.08	17.73
49	92	101	99
53.33	55.88	53.58	57.45
16.91	21.06	19.14	19.21
54	76	72	82
56.40	57.42	55.42	62.09
16.68	21.69	17.80	20.22
47	72	67	76
15.98	15.87	16.60	13.01
11.38	9.92	12.72	10.37
54	77	72	80
15.13	14.44	16.01	11.41*
11.18	9.54	12.63	11.02
47	73	67	74
4.12	8.59	3.05	3.33*
3.86	9.95	2.76	3.75
51	75	61	63
3.98	8.69	3.28	3.24*
4.07	10.48	3.28	3.67
46	72	58	62
26.04	19.53	26.35	26.44*
12.95	13.43	15.55	14.16
53	76	72	81

Manufacturing Cost Structure by Industry (*continued*)

	Industry	
Manufacturing Cost Structure	*Total*	*Consumer*
Manufacturing overhead cost (% of total mfg. cost), 1992		
Mean	23.71	22.55
Standard deviation	14.26	13.78
Number	340	64

	Industry		
Industrial Goods	*Basic*	*Machinery*	*Electronics*
25.96	19.48	25.80	25.47*
13.16	13.10	15.15	14.95
52	77	70	77

*The ANOVA F-test calculated from the comparison between industries is significant at the .05 alpha level.

(*concluded*)

Manufacturing Performance Improvement Index by Industry

	Industry	
Performance Indicators	Total	Consumer
Overall quality as perceived by customers		
Mean	113.17	109.15
Standard deviation	16.45	11.74
Number	515	85
Average unit production costs for typical product		
Mean	106.63	107.12
Standard deviation	17.01	17.78
Number	515	83
Inventory turnover		
Mean	116.76	109.73
Standard deviation	41.92	22.35
Number	514	84
Speed of new product development and/or design change		
Mean	107.19	107.20
Standard deviation	17.24	18.50
Number	509	82
On-time delivery		
Mean	112.45	108.42
Standard deviation	30.45	11.28
Number	514	84
Equipment changeover (or set-up) time		
Mean	111.65	106.91
Standard deviation	20.24	15.83
Number	498	80
Market share		
Mean	107.49	107.96
Standard deviation	18.26	14.58
Number	508	82
Profitability		
Mean	121.29	121.97
Standard deviation	57.77	37.42
Number	488	79
Customer service		
Mean	112.04	109.89
Standard deviation	16.69	16.10
Number	499	82
Manufacturing lead time		
Mean	110.25	107.83
Standard deviation	20.70	13.87
Number	512	84

Industry			
Industrial Goods	*Basic*	*Machinery*	*Electronics*
112.60	111.80	114.21	116.25*
13.08	10.69	17.20	22.36
57	116	127	130
106.86	103.06	106.05	110.02*
11.40	13.04	18.38	19.65
56	118	128	130
118.52	108.24	117.20	127.56*
28.16	15.53	27.92	71.36
56	115	128	131
104.70	104.86	108.26	109.34
17.57	13.25	17.92	18.59
57	116	126	128
111.28	109.21	114.95	116.05
16.11	13.83	51.06	27.19
57	117	127	129
111.23	108.62	113.46	115.86*
12.64	12.37	23.95	25.78
57	113	124	124
112.82	104.72	106.44	108.35
29.04	11.24	15.89	21.04
57	116	126	127
118.91	122.03	120.48	122.11
38.07	75.34	62.02	53.90
56	111	122	120
112.35	109.79	112.83	114.65
14.47	11.82	17.56	20.48
57	114	126	120
107.50	106.13	109.55	117.38*
21.17	14.02	15.65	30.02
56	115	128	129

Manufacturing Performance Improvement Index by Industry (*continued*)

	Industry	
Performance Indicators	Total	Consumer
Procurement lead time		
Mean	106.02	104.76
Standard deviation	16.64	10.85
Number	509	82
Delivery lead time		
Mean	107.88	103.38
Standard deviation	17.91	11.33
Number	510	84
Variety of products producible by manufacturing		
Mean	112.68	112.96
Standard deviation	22.38	29.07
Number	515	85

	Industry		
Industrial Goods	Basic	Machinery	Electronics
105.66	104.31	103.58	110.95*
13.57	10.37	14.43	24.79
56	116	127	128
106.79	107.20	107.17	112.77*
16.50	15.31	15.17	24.80
57	117	127	125
110.51	109.14	112.52	116.78
13.70	15.08	22.09	25.61
57	116	128	129

*The ANOVA F-test calculated from the comparison between industries is significant at the .05 alpha level.

(*concluded*)

Manufacturing Objectives by Industry

	Industry	
Manufacturing Objectives	*Total*	*Consumer*
Improve conformance quality (reduce defects)		
Mean	5.96	5.85
Standard deviation	1.05	1.06
Number	537	87
Reduce unit costs		
Mean	5.83	5.57
Standard deviation	1.10	1.32
Number	538	88
Improve safety record		
Mean	4.68	4.78
Standard deviation	1.59	1.47
Number	533	86
Reduce manufacturing lead time		
Mean	5.28	5.07
Standard deviation	1.36	1.35
Number	535	87
Increase capacity		
Mean	4.45	4.39
Standard deviation	1.68	1.73
Number	536	88
Reduce procurement lead time		
Mean	4.98	4.77
Standard deviation	1.36	1.22
Number	532	88
Reduce new product development cycle		
Mean	5.45	5.38
Standard deviation	1.36	1.37
Number	534	87
Reduce materials costs		
Mean	5.37	4.83
Standard deviation	1.24	1.39
Number	533	88
Reduce overhead costs		
Mean	5.52	5.55
Standard deviation	1.19	1.16
Number	534	87
Improve direct labor productivity		
Mean	5.40	5.31
Standard deviation	1.23	1.28
Number	534	87

	Industry		
Industrial Goods	*Basic*	*Machinery*	*Electronics*
5.93	5.90	6.19	5.86
0.96	1.22	0.90	1.03
57	126	133	134
5.70	5.70	5.95	6.08*
1.07	1.13	1.02	0.96
57	126	133	134
4.67	5.20	4.72	4.08*
1.53	1.64	1.53	1.53
57	125	132	133
4.95	4.76	5.54	5.80*
1.56	1.53	1.14	1.02
57	125	132	134
3.96	4.89	4.57	4.18*
1.74	1.65	1.64	1.60
57	125	133	133
4.49	4.33	5.35	5.58*
1.28	1.53	1.25	1.02
57	123	133	131
5.07	4.91	5.69	5.90*
1.39	1.37	1.24	1.24
57	124	133	133
5.32	5.25	5.53	5.69*
1.10	1.24	1.20	1.08
57	122	133	133
5.39	5.41	5.53	5.66
1.42	1.21	1.18	1.07
57	125	133	132
5.30	5.54	5.53	5.23
1.31	1.17	1.13	1.31
57	124	133	133

Manufacturing Objectives by Industry (*continued*)

	Industry	
Manufacturing Objectives	*Total*	*Consumer*
Increase throughput		
Mean	5.44	5.36
Standard deviation	1.21	1.25
Number	533	88
Reduce number of vendors		
Mean	4.04	4.02
Standard deviation	1.46	1.37
Number	532	87
Improve vendor quality		
Mean	5.48	5.53
Standard deviation	1.19	1.22
Number	533	88
Reduce inventories		
Mean	5.19	5.13
Standard deviation	1.36	1.44
Number	536	88
Increase delivery reliability		
Mean	5.47	5.32
Standard deviation	1.28	1.37
Number	537	88
Increase delivery speed		
Mean	5.13	4.86
Standard deviation	1.28	1.38
Number	538	88
Improve ability to make rapid product mix changes		
Mean	4.91	4.79
Standard deviation	1.34	1.38
Number	534	87
Improve ability to make rapid volume changes		
Mean	4.68	4.70
Standard deviation	1.33	1.31
Number	535	87
Reduce break-even points		
Mean	5.08	4.78
Standard deviation	1.34	1.43
Number	530	85
Raise employee morale		
Mean	5.15	5.10
Standard deviation	1.13	1.06
Number	538	88

	Industry		
Industrial Goods	*Basic*	*Machinery*	*Electronics*
5.19	5.27	5.73	5.47*
1.19	1.38	1.03	1.14
57	124	132	132
3.86	3.55	4.16	4.47*
1.30	1.41	1.56	1.41
56	123	133	133
5.38	5.02	5.73	5.67*
1.17	1.32	1.06	1.04
56	123	133	133
5.09	4.82	5.26	5.56*
1.27	1.59	1.28	1.09
56	125	133	134
5.26	5.36	5.56	5.69
1.38	1.36	1.30	1.04
57	125	133	134
4.84	4.90	5.34	5.46*
1.26	1.33	1.30	1.04
57	126	133	134
4.60	4.59	4.98	5.34*
1.35	1.40	1.22	1.24
57	125	132	133
4.49	4.39	4.77	4.93*
1.14	1.38	1.30	1.36
57	124	133	134
4.65	5.03	5.30	5.28*
1.38	1.36	1.24	1.26
57	122	132	134
5.07	5.00	5.38	5.13
0.90	1.21	1.14	1.16
57	126	133	134

Manufacturing Objectives by Industry (*continued*)

	Industry	
Manufacturing Objectives	*Total*	*Consumer*
Maximize cash flow		
Mean	5.11	4.95
Standard deviation	1.29	1.34
Number	526	86
Increase environmental safety and protection		
Mean	4.87	5.03
Standard deviation	1.46	1.34
Number	536	88
Reduce capacity		
Mean	2.25	2.24
Standard deviation	1.45	1.45
Number	534	88
Increase product or materials standardization		
Mean	4.68	4.52
Standard deviation	1.38	1.41
Number	534	88
Improve labor relations		
Mean	4.49	4.33
Standard deviation	1.38	1.47
Number	535	88
Improve white-collar productivity		
Mean	5.15	4.90
Standard deviation	1.16	1.31
Number	533	88
Increase range of products produced by existing facilities		
Mean	4.08	4.02
Standard deviation	1.53	1.65
Number	535	88
Meet financial shipping goals		
Mean	4.62	4.31
Standard deviation	1.43	1.47
Number	518	83
Improve pre-sales service and technical support		
Mean	4.54	3.83
Standard deviation	1.33	1.51
Number	522	83
Improve after-sales service		
Mean	4.69	3.99
Standard deviation	1.36	1.45
Number	525	84

	Industry		
Industrial Goods	*Basic*	*Machinery*	*Electronics*
5.21	5.02	5.30	5.09
1.44	1.30	1.19	1.28
57	123	131	129
4.91	5.48	4.73	4.32*
1.35	1.38	1.39	1.49
57	125	133	133
2.14	1.88	2.33	2.57*
1.51	1.34	1.46	1.45
57	124	132	133
4.34	4.15	4.93	5.17*
1.20	1.45	1.32	1.22
56	124	133	133
4.39	4.45	4.75	4.40
1.19	1.23	1.43	1.45
57	124	133	133
4.80	5.02	5.37	5.39*
1.18	1.14	1.00	1.13
56	124	132	133
3.72	4.18	4.17	4.11
1.53	1.61	1.53	1.38
57	124	133	133
4.65	4.30	4.66	5.06*
1.44	1.37	1.38	1.39
52	124	128	131
4.25	4.51	4.84	4.84*
1.19	1.29	1.20	1.20
56	123	132	128
4.18	4.65	5.17	4.91*
1.36	1.29	1.21	1.26
56	124	133	128

Manufacturing Objectives by Industry (*continued*)

	Industry	
Manufacturing Objectives	*Total*	*Consumer*
Change culture of manufacturing organization		
Mean	5.14	5.17
Standard deviation	1.44	1.63
Number	533	88
Improve interfunctional communication		
Mean	5.43	5.42
Standard deviation	1.13	1.25
Number	535	88
Improve communication with external partners		
Mean	5.10	4.89
Standard deviation	1.27	1.30
Number	535	88
Reduce set-up/changeover times		
Mean	5.00	4.95
Standard deviation	1.42	1.59
Number	533	88.

	Industry		
Industrial Goods	Basic	Machinery	Electronics
5.32	5.02	5.25	5.05
1.49	1.37	1.42	1.39
56	124	132	133
5.23	5.37	5.45	5.56
1.29	1.04	1.06	1.12
56	125	133	133
5.02	4.97	5.12	5.38*
1.14	1.36	1.22	1.14
57	124	133	133
4.84	4.77	5.26	5.07
1.50	1.44	1.29	1.35
56	124	132	133

*The ANOVA F-test calculated from the comparison between industries is significant at the .05 alpha level.

(*concluded*)

Past Pay-offs from Programs and Activities by Industry

	Industry	
Programs / Activities	Total	Consumer
Giving workers a broad range of tasks and/or more responsibility		
Mean	4.25	3.86
Standard deviation	1.40	1.40
Number	430	69
Activity-based costing		
Mean	3.65	3.63
Standard deviation	1.54	1.51
Number	288	40
Manufacturing reorganization		
Mean	4.58	4.52
Standard deviation	1.41	1.55
Number	401	64
Worker training		
Mean	4.55	4.31
Standard deviation	1.27	1.28
Number	428	68
Management training		
Mean	4.61	4.62
Standard deviation	1.24	1.07
Number	411	68
Supervisor training		
Mean	4.62	4.58
Standard deviation	1.21	1.07
Number	413	66
Computer-aided manufacturing (CAM)		
Mean	4.13	3.54
Standard deviation	1.65	1.68
Number	350	46
Computer-aided design (CAD)		
Mean	4.47	4.02
Standard deviation	1.67	1.77
Number	395	49
Value analysis/product redesign		
Mean	4.30	3.57
Standard deviation	1.57	1.57
Number	320	37
Interfunctional work teams		
Mean	4.48	4.55
Standard deviation	1.33	1.11
Number	373	56

	Industry		
Industrial Goods	*Basic*	*Machinery*	*Electronics*
4.37	4.09	4.55	4.29*
1.43	1.42	1.33	1.38
43	103	111	104
4.06	3.43	3.88	3.61
1.95	1.47	1.57	1.51
16	83	72	77
4.71	4.36	4.80	4.57
1.43	1.41	1.28	1.42
35	99	98	105
4.39	4.60	4.67	4.59
1.28	1.36	1.29	1.14
44	113	107	96
4.54	4.60	4.71	4.53
1.57	1.20	1.27	1.24
39	109	103	92
4.66	4.67	4.66	4.55
1.20	1.19	1.35	1.19
44	109	103	91
4.00	4.01	4.44	4.28*
1.96	1.54	1.63	1.62
25	93	96	90
4.06	3.57	5.06	4.95*
2.09	1.67	1.36	1.33
34	91	115	106
4.14	3.84	4.76	4.64*
1.91	1.46	1.35	1.60
22	85	91	85
4.55	4.44	4.65	4.26
1.52	1.40	1.30	1.35
33	97	100	87

Past Pay-offs from Programs and Activities by Industry (*continued*)

	Industry	
Programs / Activities	*Total*	*Consumer*
Quality function deployment		
Mean	4.63	4.36
Standard deviation	1.38	1.52
Number	375	56
Developing new processes for new products		
Mean	4.64	4.80
Standard deviation	1.42	1.31
Number	378	55
Developing new processes for old products		
Mean	4.58	4.54
Standard deviation	1.49	1.40
Number	360	63
Integrating information systems in manufacturing		
Mean	4.34	4.13
Standard deviation	1.40	1.47
Number	403	61
Integrating information systems across functions		
Mean	4.23	4.09
Standard deviation	1.42	1.30
Number	374	58
Reconditioning physical plants		
Mean	4.34	4.53
Standard deviation	1.49	1.54
Number	334	58
Just-in-time		
Mean	4.39	4.09
Standard deviation	1.67	1.63
Number	371	53
Robots		
Mean	3.59	2.90
Standard deviation	1.88	1.79
Number	283	40
Flexible manufacturing systems		
Mean	3.98	3.86
Standard deviation	1.58	1.42
Number	319	43
Design for manufacture		
Mean	4.04	3.44
Standard deviation	1.55	1.48
Number	308	36

	Industry		
Industrial Goods	Basic	Machinery	Electronics
4.52	4.56	4.87	4.68
1.26	1.35	1.36	1.39
31	97	98	93
4.29	4.72	4.35	4.86*
1.40	1.58	1.29	1.36
34	98	91	100
4.67	4.57	4.60	4.56
1.49	1.63	1.44	1.47
30	98	87	82
3.73	4.39	4.38	4.58*
1.42	1.34	1.47	1.29
33	102	104	103
3.69	4.31	4.27	4.33
1.49	1.47	1.38	1.46
26	93	100	97
4.09	4.34	4.52	4.07
1.60	1.41	1.33	1.63
22	89	84	81
4.31	3.90	4.51	4.84*
1.64	1.64	1.65	1.64
32	84	94	108
3.08	3.17	4.29	3.74*
1.89	1.98	1.62	1.85
13	78	80	72
3.68	3.58	4.22	4.25*
1.60	1.57	1.43	1.73
19	83	90	84
3.62	3.33	4.48	4.44*
1.43	1.55	1.51	1.39
21	66	88	97

Past Pay-offs from Programs and Activities by Industry (*continued*)

	Industry	
Programs / Activities	*Total*	*Consumer*
Statistical quality control		
Mean	4.50	4.50
Standard deviation	1.45	1.28
Number	399	60
Closing and/or relocating plants		
Mean	3.22	3.63
Standard deviation	2.14	2.14
Number	287	38
Quality circles		
Mean	4.22	4.06
Standard deviation	1.68	1.74
Number	330	47
Investing in improved production-inventory control systems		
Mean	4.30	4.17
Standard deviation	1.55	1.42
Number	357	52
Hiring in new skills from outside		
Mean	3.89	4.02
Standard deviation	1.56	1.58
Number	323	48
Linking manufacturing strategy to business strategy		
Mean	4.63	4.67
Standard deviation	1.45	1.34
Number	413	67

	Industry		
Industrial Goods	**Basic**	**Machinery**	**Electronics**
4.52	4.25	4.45	4.79
1.71	1.53	1.47	1.28
42	102	98	97
4.52	2.74	3.18	3.18*
1.89	2.11	2.14	2.10
21	76	73	79
3.61	4.08	4.32	4.51
1.72	1.70	1.54	1.74
18	92	88	85
4.00	4.32	4.38	4.34
1.75	1.49	1.56	1.63
27	96	95	87
4.05	3.56	4.11	3.87
1.60	1.48	1.54	1.61
21	85	87	82
4.70	4.65	4.51	4.66
1.38	1.55	1.51	1.39
40	100	101	105

*The ANOVA F-test calculated from the comparison between industries is significant at the .05 alpha level.

(concluded)

Future Emphasis on Programs and Activities by Industry

	Industry	
Programs / Activities	*Total*	*Consumer*
Giving workers a broad range of tasks and/or more responsibility		
Mean	5.09	4.95
Standard deviation	1.47	1.54
Number	495	80
Activity-based costing		
Mean	4.12	3.79
Standard deviation	1.69	1.75
Number	451	72
Manufacturing reorganization		
Mean	4.57	4.50
Standard deviation	1.57	1.79
Number	486	78
Worker training		
Mean	5.25	5.33
Standard deviation	1.14	1.31
Number	494	81
Management training		
Mean	5.19	5.20
Standard deviation	1.28	1.46
Number	496	83
Supervisor training		
Mean	1.19	1.18
Standard deviation	1.19	1.18
Number	490	79
Computer-aided manufacturing (CAM)		
Mean	4.53	3.99
Standard deviation	1.86	2.05
Number	474	71
Computer-aided design (CAD)		
Mean	4.72	3.97
Standard deviation	1.74	1.82
Number	490	70
Value analysis/product redesign		
Mean	4.52	3.93
Standard deviation	1.67	1.79
Number	459	68
Interfunctional work teams		
Mean	4.99	5.07
Standard deviation	1.50	1.49
Number	469	74

	Industry		
Industrial Goods	*Basic*	*Machinery*	*Electronics*
5.31	4.87	5.15	5.22
1.45	1.47	1.46	1.42
54	118	119	124
3.88	4.08	4.32	4.29
1.73	1.61	1.69	1.67
51	104	110	114
4.17	4.34	4.81	4.78*
1.72	1.64	1.32	1.46
54	115	115	124
5.34	5.30	5.29	5.06
1.05	1.04	1.07	1.21
56	117	119	121
4.87	5.24	5.24	5.23
1.50	1.13	1.20	1.24
55	119	118	121
1.27	1.07	1.15	1.32
1.27	1.07	1.15	1.32
57	118	118	118
3.78	4.56	5.02	4.66*
1.90	1.86	1.55	1.85
54	108	122	119
4.36	3.90	5.43	5.28*
1.65	1.86	1.32	1.50
53	113	129	125
4.10	3.99	5.12	4.93*
1.78	1.57	1.43	1.57
49	109	115	118
4.62	4.83	5.11	5.13
1.89	1.38	1.32	1.56
52	112	115	116

Future Emphasis on Programs and Activities by Industry (*continued*)

| Programs / Activities | Industry | |
	Total	Consumer
Quality function deployment		
Mean	5.21	4.97
Standard deviation	1.45	1.52
Number	475	71
Developing new processes for new products		
Mean	5.16	4.96
Standard deviation	1.57	1.67
Number	490	76
Developing new processes for old products		
Mean	4.54	4.71
Standard deviation	1.71	1.64
Number	478	76
Integrating information systems in manufacturing		
Mean	5.30	5.34
Standard deviation	1.33	1.49
Number	497	77
Integrating information systems across functions		
Mean	5.17	4.99
Standard deviation	1.39	1.53
Number	485	77
Reconditioning physical plants		
Mean	4.25	4.41
Standard deviation	1.67	1.61
Number	459	73
Just-in-time		
Mean	4.89	4.92
Standard deviation	1.64	1.56
Number	474	71
Robots		
Mean	3.41	2.79
Standard deviation	2.00	1.87
Number	450	68
Flexible manufacturing systems		
Mean	4.44	4.25
Standard deviation	1.85	1.83
Number	464	69
Design for manufacture		
Mean	4.72	3.88
Standard deviation	1.81	1.86
Number	445	64

Industry			
Industrial Goods	*Basic*	*Machinery*	*Electronics*
5.04	5.27	5.34	5.25
1.59	1.49	1.30	1.43
54	112	117	121
4.57	5.18	5.09	5.57*
1.82	1.72	1.40	1.29
54	114	119	127
4.70	4.93	4.47	4.04*
1.59	1.66	1.63	1.81
56	111	118	117
4.68	5.39	5.38	5.41*
1.45	1.23	1.19	1.34
57	114	125	124
4.61	5.18	5.36	5.37*
1.52	1.36	1.20	1.37
56	114	121	117
3.71	4.47	4.45	3.98*
1.57	1.52	1.70	1.78
51	108	112	115
4.65	4.40	5.14	5.18*
1.78	1.64	1.52	1.66
55	107	118	123
2.57	3.18	4.03	3.73*
1.76	1.97	2.09	1.86
49	103	117	113
3.67	3.94	5.06	4.76*
1.87	1.88	1.73	1.68
54	106	116	119
4.02	3.93	5.37	5.51*
1.83	1.73	1.47	1.53
50	98	116	117

Future Emphasis on Programs and Activities by Industry (*continued*)

	Industry	
Programs / Activities	*Total*	*Consumer*
Statistical quality control		
Mean	5.03	5.12
Standard deviation	1.50	1.33
Number	485	77
Closing and/or relocating plants		
Mean	2.89	3.31
Standard deviation	2.07	2.25
Number	435	67
Quality circles		
Mean	1.92	2.07
Standard deviation	1.92	2.07
Number	456	67
Investing in improved production-inventory control systems		
Mean	4.87	4.96
Standard deviation	1.61	1.65
Number	474	69
Hiring in new skills from outside		
Mean	4.06	3.83
Standard deviation	1.72	1.67
Number	465	72
Linking manufacturing strategy to business strategy		
Mean	5.59	5.55
Standard deviation	1.25	1.27
Number	490	80

	Industry		
Industrial Goods	*Basic*	*Machinery*	*Electronics*
5.16	5.09	4.95	4.93
1.95	1.46	1.39	1.53
56	112	117	123
2.76	2.81	2.76	2.87
2.11	2.05	1.96	2.08
50	102	109	107
2.21	1.96	1.65	1.85*
2.21	1.96	1.65	1.85
50	110	115	114
4.50	4.79	4.84	5.08
1.71	1.67	1.49	1.60
56	112	118	119
3.65	3.87	4.33	4.29*
1.58	1.82	1.59	1.78
54	108	117	114
5.58	5.61	5.54	5.65
1.10	1.36	1.28	1.16
57	114	118	121

*The ANOVA F-test calculated from the comparison between industries is significant at the .05 alpha level.

(*concluded*)

Internal Environment by Industry

| | Industry | | |
Indicators	Total	Consumer	Industrial Goods
Top management only			
Mean	0.03	0.06	0.05
Standard deviation	0.18	0.23	0.23
Number	539	88	57
Top and some middle management			
Mean	0.26	0.31	0.35
Standard deviation	0.44	0.46	0.48
Number	539	88	57
Top and most middle management			
Mean	0.31	0.34	0.21
Standard deviation	0.46	0.48	0.41
Number	539	88	57
Every manager and supervisor			
Mean	0.25	0.22	0.19
Standard deviation	0.43	0.41	0.40
Number	539	88	57
Every manager, supervisor, and worker			
Mean	0.13	0.08	0.18
Standard deviation	0.34	0.27	0.38
Number	539	88	57

Industry		
Basic	*Machinery*	*Electronics*
0.04	0.02	0.01
0.20	0.15	0.12
126	133	135
0.22	0.28	0.21
0.42	0.45	0.41
126	133	135
0.32	0.26	0.39
0.47	0.44	0.49
126	133	135
0.25	0.26	0.29
0.44	0.44	0.45
126	133	135
0.14	0.18	0.10
0.35	0.39	0.30
126	133	135

Past and Future Percentage of Domestic and Foreign Sales, Productions, and Purchases by Industry

	Industry	
Competitive Abilities	Total	Consumer
Domestic sales, 1989		
Mean	85.41	92.87
Standard deviation	59.94	12.65
Number	160	29
Foreign sales, 1989		
Mean	21.39	7.13
Standard deviation	35.61	12.65
Number	160	29
Domestic sales, 1992		
Mean	82.42	91.12
Standard deviation	69.18	13.44
Number	155	28
Foreign sales, 1992		
Mean	25.90	8.88
Standard deviation	43.43	13.44
Number	155	28
Domestic production, 1989		
Mean	89.30	91.70
Standard deviation	16.12	13.48
Number	155	29
Foreign production, 1989		
Mean	10.70	8.30
Standard deviation	16.12	13.48
Number	155	29
Domestic production, 1992		
Mean	85.94	90.19
Standard deviation	19.21	14.03
Number	152	28
Foreign production, 1992		
Mean	14.06	9.81
Standard deviation	19.21	14.03
Number	152	28
Domestic purchases, 1989		
Mean	84.25	86.64
Standard deviation	18.18	17.91
Number	153	29
Foreign purchases, 1989		
Mean	15.75	13.36
Standard deviation	18.18	17.91
Number	153	29

	Industry		
Industrial Goods	Basic	Machinery	Electronics
85.97	80.05	96.37	72.43
18.81	19.54	123.67	18.93
37	19	35	40
14.03	19.95	35.00	27.33*
18.81	19.54	65.86	19.06
37	19	35	40
82.66	78.31	94.06	66.90
19.81	20.38	139.30	20.74
36	16	36	39
17.34	21.69	42.05	32.85*
19.81	20.38	80.49	20.89
36	16	36	39
90.28	90.88	91.34	83.97
14.54	17.33	13.73	20.01
37	17	34	38
9.72	9.12	8.66	16.03
14.54	17.33	13.73	20.01
37	17	34	38
87.01	89.25	85.76	80.43
18.99	19.34	21.30	20.32
36	16	35	37
12.99	10.75	14.24	19.57
18.99	19.34	21.30	20.32
36	16	35	37
87.69	84.99	84.66	78.32
11.79	18.28	18.18	22.50
36	17	34	37
12.31	15.01	15.34	21.68
11.79	18.28	18.18	22.50
36	17	34	37

**Past and Future Percentage of Domestic and Foreign Sales,
Productions, and Purchases by Industry** (*continued*)

	Industry	
Competitive Abilities	*Total*	*Consumer*
Domestic purchases, 1992		
Mean	82.21	85.11
Standard deviation	17.88	17.70
Number	160	29
Foreign purchases, 1992		
Mean	17.79	14.89
Standard deviation	17.88	17.70
Number	160	29

	Industry		
Industrial Goods	**Basic**	**Machinery**	**Electronics**
85.97	85.37	82.13	74.97
14.21	16.23	16.29	21.75
37	19	35	40
14.03	14.63	17.87	25.03
14.21	16.23	16.29	21.75
37	19	35	40

*The ANOVA F-test calculated from the comparison between industries within the United States is significant at the .05 alpha level.

(*concluded*)

APPENDIX C

COUNTRY BY INDUSTRY DATA

Business Unit Growth Strategy by Industry (U.S.)

| | Industry | | |
Indicators	Total	Consumer	Industrial Goods
Build market share			
Mean	0.79	0.81	0.87
Standard deviation	0.41	0.40	0.34
Number	182	32	39
Hold (defend) market share			
Mean	0.18	0.16	0.10
Standard deviation	0.38	0.37	0.31
Number	182	32	39
Harvest (maximize cash flow, sacrifice market share)			
Mean	0.03	0.03	0.03
Standard deviation	0.18	0.18	0.16
Number	182	32	39
Withdraw (prepare to exit the business)			
Mean	0.01	0.00	0.00
Standard deviation	0.07	0.00	0.00
Number	182	32	39

	Industry	
Basic	Machinery	Electronics
0.75	0.77	0.73
0.44	0.43	0.45
24	43	44
0.25	0.16	0.23
0.44	0.37	0.42
24	43	44
0.00	0.07	0.02
0.00	0.26	0.15
24	43	44
0.00	0.00	0.02
0.00	0.00	0.15
24	43	44

Selected Business Unit Profile by Industry (U.S.)

Indicators	Industry	
	Total	Consumer
Annual sales revenue[†]		
Mean	1,584.30	762.82
Standard deviation	6,177.42	983.07
Number	163	27
Pre-tax return on assets (% of assets)		
Mean	16.67	20.15
Standard deviation	15.05	12.49
Number	119	21
Net pre-tax profit ratio (% of sales)		
Mean	9.75	13.84
Standard deviation	9.54	11.05
Number	122	20
R&D expenses (% of sales)		
Mean	4.36	3.02
Standard deviation	4.58	4.25
Number	146	27
Growth rate in unit sales (%)		
Mean	7.61	6.10
Standard deviation	10.63	6.33
Number	156	27
Market share of primary product		
Mean	38.65	42.48
Standard deviation	22.32	25.10
Number	145	25
Capacity utilization (%)		
Mean	72.28	72.53
Standard deviation	18.30	16.04
Number	155	27
Number of plants in the business unit		
Mean	10.59	13.73
Standard deviation	16.57	12.42
Number	175	30
Number of employees		
Mean	8,747.12	5,244.10
Standard deviation	32,636.78	6,842.45
Number	175	31
Manufacturing direct labor employees		
Mean	2,469.11	2,596.11
Standard deviation	6,230.05	3,667.05
Number	156	27

	Industry		
Industrial Goods	*Basic*	*Machinery*	*Electronics*
309.22	960.41	1,476.85	3,664.00
438.77	1,543.72	3,876.20	11,493.34
36	21	38	41
18.32	16.81	12.07	16.18
18.38	11.80	12.58	15.99
32	14	25	27
8.37	8.39	8.11	10.57
6.73	5.78	9.40	11.93
32	14	26	30
2.29	3.01	4.25	7.87*
2.18	3.14	3.33	5.78
34	17	30	38
10.36	1.51	7.21	9.80*
10.74	7.67	12.37	11.39
35	20	37	37
34.43	34.83	41.15	39.24
26.79	19.74	21.49	24.62
32	18	35	35
72.36	78.48	67.79	72.51
19.60	23.80	19.31	13.23
37	21	33	37
7.38	15.61	5.90	13.17
8.98	20.05	8.77	25.12
39	23	41	42
2,041.92	5,325.23	8,089.63	19,993.19
3,165.72	7,169.41	18,184.82	62,766.38
39	22	41	42
954.03	1,451.00	2,199.95	4,506.05
1,802.11	1,732.72	3,503.94	11,369.21
35	17	39	38

Selected Business Unit Profile by Industry (U.S.) (*continued*)

Indicators	Industry	
	Total	*Consumer*
Manufacturing indirect labor employees		
Mean	1,378.47	759.74
Standard deviation	3,862.31	1,117.27
Number	154	27

		Industry		
	Industrial Goods	**Basic**	**Machinery**	**Electronics**
	393.00	1,058.94	1,530.08	2,649.31
	646.29	2,211.47	3,822.40	6,310.66
	34	16	38	39

†In U.S. dollars.

*The ANOVA F-test calculated from the comparison between industries within the United States is significant at the .05 alpha level.

(concluded)

Importance of Competitive Abilities by Industry (U.S.)

	Industry	
Importance of Competitive Abilities	Total	Consumer
Ability to profit in price-competitive market		
Mean	5.73	5.69
Standard deviation	1.28	1.12
Number	182	32
Ability to make rapid changes in design		
Mean	5.17	4.59
Standard deviation	1.37	1.52
Number	182	32
Ability to introduce new products quickly		
Mean	5.48	5.41
Standard deviation	1.19	1.07
Number	181	32
Ability to make rapid volume changes		
Mean	5.05	5.13
Standard deviation	1.40	1.31
Number	181	32
Ability to make rapid product mix changes		
Mean	5.37	5.06
Standard deviation	1.17	1.05
Number	182	32
Ability to offer a broad product line		
Mean	5.36	5.31
Standard deviation	1.33	1.23
Number	180	32
Ability to offer consistently low defect rates		
Mean	6.51	6.28
Standard deviation	0.73	0.81
Number	182	32
Ability to provide high-performance products or product amenities		
Mean	6.00	5.50
Standard deviation	1.02	1.48
Number	182	32
Ability to provide reliable/durable products		
Mean	6.31	6.00
Standard deviation	0.93	0.95
Number	180	32
Ability to provide fast deliveries		
Mean	5.69	6.00
Standard deviation	1.05	0.92
Number	182	32

	Industry		
Industrial Goods	*Basic*	*Machinery*	*Electronics*
5.54	5.54	5.69	6.04
1.45	1.25	1.39	1.14
39	24	43	44
5.15	4.38	5.56	5.66*
1.39	1.47	1.16	1.03
39	24	43	44
5.15	4.88	5.64	6.00*
1.29	1.33	1.12	0.92
39	24	42	44
5.39	4.50	4.95	5.09
1.24	1.69	1.33	1.43
38	24	43	44
5.51	5.42	5.33	5.50
1.32	1.10	1.15	1.17
39	24	43	44
5.41	5.22	5.53	5.25
1.27	1.27	1.47	1.37
39	22	43	44
6.67	6.33	5.56	5.57
0.58	1.01	0.70	0.63
39	24	43	44
6.05	6.88	6.09	6.30*
0.94	0.80	0.95	0.73
39	24	43	44
6.33	5.78	6.60	6.51*
0.90	1.38	0.66	0.70
39	23	43	43
5.74	4.46	5.56	5.66
1.09	0.93	1.08	1.12
39	24	43	44

Importance of Competitive Abilities by Industry (U.S.) (*continued*)

Importance of Competitive Abilities	*Industry*	
	Total	*Consumer*
Ability to make dependable delivery promises		
Mean	6.34	6.44
Standard deviation	0.75	0.56
Number	182	32
Ability to provide effective after-sales service		
Mean	5.47	5.03
Standard deviation	1.34	1.64
Number	181	31
Ability to provide product support effectively		
Mean	5.65	5.16
Standard deviation	1.20	1.46
Number	182	32
Ability to make products easily available (broad distribution)		
Mean	5.54	6.06
Standard deviation	1.29	1.01
Number	181	32
Ability to customize products and services to customer needs		
Mean	5.47	5.13
Standard deviation	1.21	1.31
Number	182	32

Industry			
Industrial Goods	*Basic*	*Machinery*	*Electronics*
6.31	6.38	6.23	6.36
0.89	0.77	0.75	0.75
39	24	43	44
5.18	5.21	5.81	5.84*
1.35	1.14	1.35	1.03
39	24	43	44
5.38	5.33	6.14	5.93*
1.25	1.46	0.83	0.85
39	24	43	44
5.58	5.25	5.23	5.59
1.33	1.36	1.46	1.13
38	24	43	44
5.38	5.25	5.70	5.70
1.21	1.22	1.21	1.09
39	24	43	44

*The ANOVA F-test calculated from the comparison between industries within the United States is significant at the .05 alpha level.

(*concluded*)

Strength in Competitive Abilities by Industry (U.S.)

Competitive Abilities	Industry	
	Total	Consumer
Ability to profit in price-competitive market		
Mean	4.55	4.90
Standard deviation	1.39	1.33
Number	182	32
Ability to make rapid changes in design		
Mean	4.46	4.16
Standard deviation	1.27	1.46
Number	182	32
Ability to introduce new products quickly		
Mean	4.42	4.25
Standard deviation	1.27	1.50
Number	181	32
Ability to make rapid volume changes		
Mean	4.88	5.09
Standard deviation	1.28	1.42
Number	182	32
Ability to make rapid product mix changes		
Mean	5.00	5.00
Standard deviation	1.12	1.22
Number	181	32
Ability to offer a broad product line		
Mean	5.16	5.16
Standard deviation	1.37	1.67
Number	180	32
Ability to offer consistently low defect rates		
Mean	5.51	5.72
Standard deviation	1.11	0.99
Number	182	32
Ability to provide high-performance products or product amenities		
Mean	5.49	5.34
Standard deviation	1.16	1.29
Number	182	32
Ability to provide reliable/durable products		
Mean	5.80	5.75
Standard deviation	0.94	0.80
Number	181	32
Ability to provide fast deliveries		
Mean	5.18	5.25
Standard deviation	1.10	1.08
Number	182	32

Industry			
Industrial Goods	*Basic*	*Machinery*	*Electronics*
4.84	4.75	4.19	4.27
1.57	1.36	1.18	1.40
39	24	43	44
4.90	4.29	4.30	4.54
1.21	1.33	1.32	1.02
39	24	43	44
4.62	4.63	4.50	4.18
1.31	1.35	1.15	1.13
39	24	42	44
5.07	4.38	4.77	4.95
1.35	1.58	1.15	1.01
39	24	43	44
5.05	4.79	5.00	5.06
1.34	1.25	1.02	0.88
39	24	43	43
5.18	5.05	5.14	5.20
1.34	1.59	1.34	1.09
39	22	43	44
5.44	5.08	5.33	5.82*
1.25	1.13	1.06	1.02
39	24	43	44
5.67	5.21	5.37	5.73
1.15	1.01	1.22	1.04
39	24	43	44
6.03	5.43	5.86	5.75
0.84	1.38	0.92	0.84
39	23	43	44
5.41	5.33	4.84	5.16
1.19	1.01	1.15	1.03
39	24	43	44

Strength in Competitive Abilities by Industry (U.S.) (*continued*)

	Industry	
Competitive Abilities	*Total*	*Consumer*
Ability to make dependable delivery promises		
Mean	5.39	5.44
Standard deviation	1.24	1.32
Number	181	32
Ability to provide effective after-sales service		
Mean	5.43	5.45
Standard deviation	1.14	1.31
Number	181	31
Ability to provide product support effectively		
Mean	5.42	5.31
Standard deviation	1.06	10.20
Number	182	32
Ability to make products easily available (broad distribution)		
Mean	5.38	5.53
Standard deviation	1.30	1.39
Number	181	32
Ability to customize products and services to customer needs		
Mean	5.13	4.88
Standard deviation	1.15	1.18
Number	182	32

	Industry		
Industrial Goods	*Basic*	*Machinery*	*Electronics*
5.69	5.22	5.30	5.27
1.03	1.28	1.35	1.23
39	23	43	44
5.33	5.25	5.65	5.36
1.15	1.03	1.15	1.06
39	2	43	44
5.51	5.25	5.58	5.34
1.17	1.07	1.00	0.89
39	24	43	44
5.55	5.04	5.40	5.27
1.29	1.46	1.31	1.13
38	24	43	44
5.59	4.96	5.14	5.00
1.01	1.00	1.30	1.10
39	24	43	44

*The ANOVA F-test calculated from the comparison between industries within the United States is significant at the .05 alpha level.

(*concluded*)

Manufacturing Cost Structure by Industry (U.S.)

Manufacturing Cost Structure	Industry	
	Total	Consumer
Manufacturing costs as percent of sales, 1989		
Mean	58.67	49.72
Standard deviation	20.54	23.21
Number	154	29
Manufacturing costs as percent of sales, 1992		
Mean	56.01	48.41
Standard deviation	18.93	23.07
Number	140	27
Materials cost (% of total mfg. cost), 1989		
Mean	56.56	54.33
Standard deviation	16.78	14.34
Number	156	26
Materials cost (% of total mfg. cost), 1992		
Mean	58.35	55.57
Standard deviation	17.34	13.23
Number	139	24
Direct labor cost (% of total mfg. cost), 1989		
Mean	12.48	15.91
Standard deviation	8.48	12.02
Number	154	26
Direct labor cost (% of total mfg. cost), 1992		
Mean	11.70	16.40
Standard deviation	8.47	12.20
Number	137	24
Energy cost (% of total mfg. cost), 1989		
Mean	4.70	4.44
Standard deviation	4.85	3.63
Number	122	23
Energy cost (% of total mfg. cost), 1992		
Mean	4.86	4.63
Standard deviation	5.31	3.66
Number	112	21
Manufacturing overhead cost (% of total mfg. cost), 1989		
Mean	27.48	25.83
Standard deviation	13.51	11.51
Number	156	26

	Industry		
Industrial Goods	Basic	Machinery	Electronics
63.42	67.61	63.09	52.34*
19.02	16.63	21.44	16.78
37	18	34	36
59.94	69.21	58.44	48.88*
18.07	9.43	17.99	15.35
37	15	31	30
57.40	50.62	58.04	58.68*
17.23	19.14	15.32	18.02
37	18	36	38
59.74	51.22	58.52	62.21
17.24	20.54	15.10	19.91
34	16	32	33
13.07	15.67	10.38	9.87*
9.21	8.60	5.24	5.65
38	18	36	36
11.81	14.63	9.68	8.50*
8.42	9.19	5.17	5.13
34	16	32	31
4.23	8.43	3.08	4.65
4.11	8.77	3.07	3.19
35	17	25	22
4.17	8.84	3.50	4.60
4.40	9.65	3.93	3.28
32	16	23	20
25.62	25.69	29.67	29.25*
13.78	15.92	13.12	13.84
38	18	36	38

Manufacturing Cost Structure by Industry (U.S.) (*continued*)

	Industry	
Manufacturing Cost Structure	*Total*	*Consumer*
Manufacturing overhead cost (% of total mfg. cost), 1992		
Mean	26.14	23.97
Standard deviation	14.06	11.38
Number	139	24

	Industry		
Industrial Goods	**Basic**	**Machinery**	**Electronics**
24.57	25.38	29.30	26.65
14.41	16.88	12.86	15.31
34	16	32	33

*The ANOVA F-test calculated from the comparison between industries within the United States is significant at the .05 alpha level.

(*concluded*)

Manufacturing Performance Improvement Index by Industry (U.S.)

Performance Indicators	Industry	
	Total	Consumer
Overall quality as perceived by customers		
Mean	114.95	109.26
Standard deviation	17.73	9.87
Number	179	31
Average unit production costs for typical product		
Mean	105.88	101.84
Standard deviation	17.23	11.80
Number	179	31
Inventory turnover		
Mean	121.16	105.58
Standard deviation	60.97	19.24
Number	179	31
Speed of new product development and/or design change		
Mean	105.59	102.84
Standard deviation	16.18	10.88
Number	179	31
On-time delivery		
Mean	112.46	107.29
Standard deviation	20.05	10.80
Number	179	31
Equipment changeover (or set-up) time		
Mean	111.71	108.43
Standard deviation	20.31	8.86
Number	175	30
Market share		
Mean	106.19	103.58
Standard deviation	20.45	10.19
Number	179	31
Profitability		
Mean	115.39	118.00
Standard deviation	51.53	28.34
Number	175	30
Customer service		
Mean	111.83	105.97
Standard deviation	18.22	10.83
Number	179	31
Manufacturing lead time		
Mean	108.29	103.19
Standard deviation	24.99	11.41
Number	179	31

	Industry		
Industrial Goods	Basic	Machinery	Electronics
113.28	115.67	113.54	121.36*
14.68	10.86	19.37	23.67
39	24	41	44
107.64	103.67	104.02	110.11
11.46	14.95	12.34	26.75
39	24	41	44
119.21	105.17	115.68	147.68*
30.03	9.35	15.88	113.93
39	24	41	44
102.95	104.83	107.95	108.07
19.05	15.06	12.08	20.00
39	24	41	44
111.62	107.38	111.83	120.23*
17.14	15.81	19.65	27.11
39	24	41	44
112.95	104.42	107.20	121.28*
12.42	6.28	11.16	35.42
39	24	40	42
115.79	101.42	106.00	102.30*
33.44	9.75	14.07	17.84
39	24	41	44
120.00	103.67	122.98	108.64
38.39	25.76	88.69	33.42
38	24	41	42
113.18	107.04	113.27	116.02
16.05	13.49	17.92	24.70
39	24	41	44
109.87	96.83	106.66	118.25*
23.29	11.50	20.98	36.55
39	24	41	44

**Manufacturing Performance Improvement Index by Industry
(U.S.) (*continued*)**

	Industry	
Performance Indicators	*Total*	*Consumer*
Procurement lead time		
Mean	104.63	102.90
Standard deviation	20.11	11.29
Number	179	31
Delivery lead time		
Mean	104.85	100.03
Standard deviation	19.66	6.95
Number	179	31
Variety of products producible by manufacturing		
Mean	111.28	108.29
Standard deviation	18.89	8.22
Number	179	31

	Industry		
Industrial Goods	*Basic*	*Machinery*	*Electronics*
105.64	99.63	98.54	113.36*
13.92	6.88	17.51	31.19
39	24	41	44
107.56	99.63	104.02	109.45
17.12	13.64	13.52	31.25
39	24	41	44
109.46	109.33	115.12	112.50
14.30	17.03	28.84	17.10
39	24	41	44

*The ANOVA F-test calculated from the comparison between industries within the United States is significant at the .05 alpha level.

(*concluded*)

Manufacturing Objectives by Industry (U.S)

Manufacturing's Objectives	Industry	
	Total	Consumer
Improve conformance quality (reduce defects)		
Mean	6.04	5.97
Standard deviation	0.98	1.06
Number	182	32
Reduce unit costs		
Mean	5.71	5.63
Standard deviation	1.11	1.24
Number	182	32
Improve safety record		
Mean	4.38	4.94
Standard deviation	1.60	1.64
Number	182	32
Reduce manufacturing lead time		
Mean	5.15	4.75
Standard deviation	1.35	1.41
Number	182	32
Increase capacity		
Mean	3.80	3.78
Standard deviation	1.68	1.86
Number	182	32
Reduce procurement lead time		
Mean	4.91	4.66
Standard deviation	1.32	1.04
Number	182	32
Reduce new product development cycle		
Mean	5.55	5.53
Standard deviation	1.24	0.95
Number	182	32
Reduce materials costs		
Mean	5.48	5.06
Standard deviation	1.16	1.34
Number	182	32
Reduce overhead costs		
Mean	5.59	5.53
Standard deviation	1.14	1.11
Number	182	32
Improve direct labor productivity		
Mean	4.96	4.75
Standard deviation	1.33	1.30
Number	182	32

	Industry		
Industrial Goods	Basic	Machinery	Electronics
6.05	6.08	6.23	5.86
0.89	1.10	0.92	0.98
39	24	43	44
5.51	5.63	5.67	6.05
1.05	1.01	1.19	0.99
39	24	43	44
4.77	5.38	3.95	3.52*
1.48	1.47	1.41	1.39
39	24	43	44
5.03	4.50	5.31	5.77*
1.40	1.29	1.33	1.01
39	24	43	44
4.08	4.39	3.79	3.27
1.69	1.72	1.66	1.44
39	24	43	44
4.51	4.13	5.28	5.51*
1.23	1.46	1.53	0.88
39	24	43	44
5.18	4.92	5.67	6.14*
1.34	1.41	1.15	1.07
39	24	43	44
5.49	5.08	5.70	5.77*
1.00	1.14	1.10	1.14
39	24	43	44
5.41	5.04	5.79	5.88*
1.31	1.20	0.94	1.03
39	24	43	44
5.13	5.29	5.16	4.59
1.34	1.23	1.33	1.35
39	24	43	44

Manufacturing Objectives by Industry (U.S) (*continued*)

	Industry	
Manufacturing's Objectives	Total	Consumer
Increase throughput		
Mean	5.31	5.13
Standard deviation	1.26	1.43
Number	182	32
Reduce number of vendors		
Mean	4.44	4.03
Standard deviation	1.39	1.58
Number	182	32
Improve vendor quality		
Mean	5.77	5.50
Standard deviation	1.09	1.27
Number	182	32
Reduce inventories		
Mean	5.31	5.03
Standard deviation	1.25	1.53
Number	182	32
Increase delivery reliability		
Mean	5.31	5.03
Standard deviation	1.25	1.36
Number	182	32
Increase delivery speed		
Mean	4.92	4.66
Standard deviation	1.34	1.52
Number	182	32
Improve ability to make rapid product mix changes		
Mean	4.68	4.66
Standard deviation	1.34	1.23
Number	182	32
Improve ability to make rapid volume changes		
Mean	4.45	4.50
Standard deviation	1.37	1.37
Number	182	32
Reduce break-even points		
Mean	4.93	4.68
Standard deviation	1.25	1.49
Number	182	32
Raise employee morale		
Mean	5.07	5.03
Standard deviation	1.09	1.06
Number	182	32

	Industry		
Industrial Goods	*Basic*	*Machinery*	*Electronics*
5.13	5.08	5.72	5.32
1.10	1.41	1.18	1.20
39	24	43	44
3.87	4.08	4.86	5.00*
1.24	1.10	1.34	1.26
39	24	43	44
5.59	5.58	6.07	5.95
1.12	1.14	1.01	0.89
39	24	43	44
5.10	4.96	5.56	5.64
1.07	1.40	1.18	1.04
39	24	43	44
5.26	5.38	5.26	5.59
1.21	1.01	1.48	1.06
39	24	43	44
4.85	4.67	4.93	5.32
1.27	1.40	1.47	1.01
39	24	43	44
4.54	4.29	4.58	5.14
1.31	1.20	1.45	1.34
39	24	43	44
4.51	3.67	4.30	4.91*
1.05	1.13	1.42	1.52
39	24	43	44
4.87	4.92	4.86	5.22
1.26	0.97	1.18	1.24
39	24	43	44
5.15	4.92	4.98	5.18
0.84	1.21	1.22	1.11
39	24	43	44

Manufacturing Objectives by Industry (U.S) (*continued*)

	Industry	
Manufacturing's Objectives	Total	Consumer
Maximize cash flow		
Mean	5.09	4.75
Standard deviation	1.45	1.52
Number	182	32
Increase environmental safety and protection		
Mean	4.68	5.09
Standard deviation	1.49	1.38
Number	182	32
Reduce capacity		
Mean	2.42	2.44
Standard deviation	1.59	1.54
Number	182	32
Increase product or materials standardization		
Mean	4.52	4.16
Standard deviation	1.34	1.39
Number	182	32
Improve labor relations		
Mean	4.18	3.94
Standard deviation	1.42	1.41
Number	182	32
Improve white-collar productivity		
Mean	4.96	4.75
Standard deviation	1.13	1.30
Number	182	32
Increase range of products produced by existing facilities		
Mean	4.05	4.13
Standard deviation	1.57	1.83
Number	182	32
Meet financial shipping goals		
Mean	4.84	4.22
Standard deviation	1.39	1.58
Number	182	32
Improve pre-sales service and technical support		
Mean	4.16	3.35
Standard deviation	1.27	1.40
Number	182	32
Improve after-sales service		
Mean	4.40	3.69
Standard deviation	1.36	1.33
Number	182	32

	Industry		
Industrial Goods	*Basic*	*Machinery*	*Electronics*
5.21	5.42	5.20	4.95
1.34	1.18	1.18	1.50
39	24	43	44
4.74	5.38	4.40	4.20*
1.41	1.58	1.38	1.52
39	24	43	44
2.08	1.75	2.53	2.95*
1.53	1.07	1.70	1.66
39	24	43	44
4.24	3.83	4.70	5.23*
1.02	1.27	1.47	1.10
39	24	43	44
4.38	4.71	4.14	3.93
1.16	1.08	1.61	1.56
39	24	43	44
4.72	4.83	4.97	5.38*
1.12	1.17	1.03	0.99
39	24	43	44
3.72	4.21	4.09	4.18
1.43	1.56	1.60	1.48
39	24	43	44
4.81	4.50	4.95	5.40*
1.39	1.35	1.33	1.14
39	24	43	44
4.28	4.17	4.12	4.64*
0.90	1.31	1.24	1.22
39	24	43	44
4.31	4.13	4.81	4.75*
1.14	1.36	1.47	1.24
39	24	43	44

Manufacturing Objectives by Industry (U.S) (*continued*)

	Industry	
Manufacturing's Objectives	*Total*	*Consumer*
Change culture of manufacturing organization		
Mean	5.30	5.28
Standard deviation	1.55	1.73
Number	182	32
Improve interfunctional communication		
Mean	5.52	5.28
Standard deviation	1.20	1.35
Number	182	32
Improve communication with external partners		
Mean	5.13	4.81
Standard deviation	1.39	1.47
Number	182	32
Reduce set-up/changeover times		
Mean	4.78	4.91
Standard deviation	1.38	1.47
Number	182	32

Industry			
Industrial Goods	**Basic**	**Machinery**	**Electronics**
5.29	5.33	5.37	5.23
1.41	1.17	1.76	1.54
39	24	43	44
5.26	5.50	5.51	5.95
1.29	1.02	1.28	0.89
39	24	43	44
5.08	5.13	4.91	5.61
1.38	1.10	1.66	1.08
39	24	43	44
4.74	3.96	5.00	4.95*
1.33	1.08	1.38	1.38
39	24	43	44

*The ANOVA F-test calculated from the comparison between industries within the United States is significant at the .05 alpha level.

(*concluded*)

Past Pay-offs from Programs and Activities by Industry (U.S.)

Programs / Activities	Industry	
	Total	Consumer
Giving workers a broad range of tasks and/or more responsibility		
Mean	4.35	3.64
Standard deviation	1.43	1.39
Number	136	25
Activity-based costing		
Mean	3.81	3.30
Standard deviation	1.50	1.41
Number	47	10
Manufacturing reorganization		
Mean	4.92	4.63
Standard deviation	1.34	1.67
Number	107	19
Worker training		
Mean	4.69	4.22
Standard deviation	1.33	1.20
Number	129	23
Management training		
Mean	4.57	4.43
Standard deviation	1.26	0.84
Number	112	23
Supervisor training		
Mean	4.72	4.35
Standard deviation	1.22	1.11
Number	123	23
Computer-aided manufacturing (CAM)		
Mean	4.55	3.60
Standard deviation	1.48	1.71
Number	76	10
Computer-aided design (CAD)		
Mean	4.79	4.78
Standard deviation	1.38	1.63
Number	119	18
Value analysis/product redesign		
Mean	4.68	4.13
Standard deviation	1.58	1.36
Number	65	8
Interfunctional work teams		
Mean	4.90	4.42
Standard deviation	1.34	1.14
Number	108	24

	Industry		
Industrial Goods	Basic	Machinery	Electronics
4.24	4.67	4.59	4.57
1.48	1.61	1.50	1.14
29	18	34	30
4.71	3.13	3.75	4.30
1.50	1.46	1.54	1.34
7	8	12	10
4.86	4.64	5.04	5.17
1.28	1.45	1.00	1.36
21	14	24	29
4.36	5.10	4.83	4.96
1.34	1.37	1.42	1.16
28	21	30	27
4.68	4.53	4.74	4.45
1.38	1.39	1.36	1.37
24	19	23	22
4.75	4.71	4.76	5.04
1.11	1.23	1.50	1.00
28	21	29	22
5.08	4.78	4.52	4.62
1.44	1.48	1.53	1.24
13	9	23	21
4.88	4.42	4.76	4.90
1.56	1.44	1.33	1.56
25	12	34	30
5.23	3.50	4.42	5.35*
1.24	1.60	1.77	1.32
13	8	19	17
5.10	5.31	4.89	5.00
1.22	1.55	1.57	1.15
21	13	28	22

Past Pay-offs from Programs and Activities by Industry (U.S.) (*continued*)

	Industry	
Programs / Activities	Total	Consumer
Quality function deployment		
Mean	4.61	4.18
Standard deviation	1.44	1.63
Number	90	17
Developing new processes for new products		
Mean	4.71	4.69
Standard deviation	1.28	1.25
Number	96	16
Developing new processes for old products		
Mean	4.66	4.11
Standard deviation	1.35	1.37
Number	83	18
Integrating information systems in manufacturing		
Mean	4.04	3.95
Standard deviation	1.45	1.36
Number	110	20
Integrating information systems across functions		
Mean	4.14	4.10
Standard deviation	1.39	1.26
Number	94	21
Reconditioning physical plants		
Mean	4.77	4.81
Standard deviation	1.44	1.42
Number	75	16
Just-in-time		
Mean	4.83	4.00
Standard deviation	1.59	1.26
Number	100	16
Robots		
Mean	3.33	3.55
Standard deviation	1.65	1.81
Number	42	11
Flexible manufacturing systems		
Mean	4.13	3.67
Standard deviation	1.49	1.61
Number	61	12
Design for manufacture		
Mean	4.46	4.33
Standard deviation	1.61	1.12
Number	80	9

	Industry		
Industrial Goods	*Basic*	*Machinery*	*Electronics*
5.00	4.13	4.67	4.71
0.97	1.46	1.52	1.55
20	8	24	21
4.52	4.93	4.11	5.20
1.47	1.32	1.37	0.82
23	14	18	25
5.00	5.00	4.24	5.06
1.37	1.08	1.56	1.00
19	13	17	16
3.91	4.00	3.95	4.30
1.54	1.45	1.65	1.30
22	17	24	27
4.29	3.75	4.36	4.05
1.44	1.48	1.50	1.36
14	12	25	22
4.29	4.93	5.11	4.58
1.64	1.03	1.28	1.88
14	15	18	12
4.55	4.38	4.83	5.52*
1.61	1.51	1.77	1.39
20	8	23	33
3.00	2.40	3.11	3.91
1.79	1.14	1.69	1.58
6	5	9	11
4.00	3.86	3.94	4.88
1.56	1.07	1.65	1.20
10	7	16	16
3.82	4.00	4.64	4.65
1.66	2.31	1.92	1.43
11	4	22	34

Past Pay-offs from Programs and Activities by Industry (U.S.) (*continued*)

Programs / Activities	Industry	
	Total	Consumer
Statistical quality control		
Mean	4.88	4.74
Standard deviation	1.54	1.42
Number	127	23
Closing and/or relocating plants		
Mean	4.62	4.60
Standard deviation	1.98	2.59
Number	61	10
Quality circles		
Mean	4.31	3.69
Standard deviation	1.65	1.80
Number	65	13
Investing in improved production-inventory control systems		
Mean	4.24	3.95
Standard deviation	1.63	1.39
Number	89	20
Hiring in new skills from outside		
Mean	4.29	3.64
Standard deviation	1.55	1.36
Number	68	11
Linking manufacturing strategy to business strategy		
Mean	4.85	4.63
Standard deviation	1.35	1.31
Number	133	27

Industry			
Industrial Goods	*Basic*	*Machinery*	*Electronics*
4.70	4.94	4.75	5.28
1.64	1.56	1.84	1.19
30	17	28	29
4.83	3.71	4.73	4.76
1.27	2.14	2.45	1.76
12	7	11	21
4.30	4.44	4.06	5.07
1.49	1.74	1.70	1.39
10	9	18	15
4.32	4.17	4.57	4.12
1.67	1.47	1.60	2.06
19	12	21	17
4.46	4.18	4.50	4.47
1.66	1.54	1.63	1.55
13	11	16	17
4.97	4.44	5.00	5.00
1.33	1.41	1.46	1.26
30	16	29	31

*The ANOVA F-test calculated from the comparison between industries within the United States is significant at the .05 alpha level.

(*concluded*)

Future Emphasis on Programs and Activities by Industry (U.S.)

	Industry	
Programs / Activities	Total	Consumer
Giving workers a broad range of tasks and/or more responsibility		
Mean	5.46	5.13
Standard deviation	1.27	1.34
Number	173	31
Activity-based costing		
Mean	3.87	3.30
Standard deviation	1.73	1.81
Number	165	27
Manufacturing reorganization		
Mean	4.27	4.10
Standard deviation	1.68	1.90
Number	173	29
Worker training		
Mean	5.41	5.53
Standard deviation	1.15	1.36
Number	175	30
Management training		
Mean	5.12	5.31
Standard deviation	1.37	1.35
Number	174	32
Supervisor training		
Mean	5.23	5.34
Standard deviation	1.26	1.14
Number	174	29
Computer-aided manufacturing (CAM)		
Mean	4.16	3.29
Standard deviation	1.81	1.12
Number	171	28
Computer-aided design (CAD)		
Mean	4.63	3.96
Standard deviation	1.62	1.75
Number	175	28
Value analysis/product redesign		
Mean	4.27	3.67
Standard deviation	1.70	1.62
Number	164	27
Interfunctional work teams		
Mean	5.19	5.03
Standard deviation	1.61	1.43
Number	171	31

	Industry		
Industrial Goods	Basic	Machinery	Electronics
5.56	5.58	5.53	5.50
1.16	1.25	1.43	1.19
36	24	40	42
3.97	4.20	4.02	3.85
1.70	1.77	1.62	1.77
36	20	41	41
3.89	4.35	4.56	4.40
1.75	1.99	1.40	1.52
38	23	41	42
5.34	5.46	5.55	5.23
1.12	0.93	1.22	1.07
38	24	40	43
4.92	5.08	5.28	5.02
1.46	1.21	1.47	1.31
38	24	39	41
5.33	5.29	5.32	4.95
1.22	0.91	1.35	1.45
39	24	41	41
3.86	4.19	4.70	4.45*
1.95	1.94	1.54	1.42
37	21	43	42
4.50	3.78	5.16	5.09*
1.70	1.93	1.31	1.23
38	23	43	43
4.31	3.57	4.76	4.53*
1.84	1.57	1.55	1.69
35	21	41	40
4.49	5.45	5.45	5.51*
2.09	1.26	1.31	1.58
35	22	42	41

Future Emphasis on Programs and Activities by Industry
(U.S.) (*continued*)

| | Industry | |
Programs / Activities	Total	Consumer
Quality function deployment		
Mean	5.02	4.46
Standard deviation	1.47	1.56
Number	169	26
Developing new processes for new products		
Mean	4.98	4.59
Standard deviation	1.60	1.76
Number	171	29
Developing new processes for old products		
Mean	4.11	4.30
Standard deviation	1.66	1.68
Number	171	27
Integrating information systems in manufacturing		
Mean	5.00	4.76
Standard deviation	1.35	1.79
Number	175	29
Integrating information systems across functions		
Mean	4.99	4.60
Standard deviation	1.44	1.83
Number	175	30
Reconditioning physical plants		
Mean	3.68	4.08
Standard deviation	1.66	1.50
Number	164	25
Just-in-time		
Mean	4.91	4.54
Standard deviation	1.66	1.63
Number	167	26
Robots		
Mean	2.54	2.38
Standard deviation	1.50	1.55
Number	161	26
Flexible manufacturing systems		
Mean	4.03	3.65
Standard deviation	1.78	1.96
Number	165	26
Design for manufacture		
Mean	4.95	3.89
Standard deviation	1.73	1.83
Number	165	27

	Industry		
Industrial Goods	Basic	Machinery	Electronics
5.11	4.86	5.90	5.10
1.64	1.52	1.25	1.38
38	22	41	42
4.76	5.24	4.76	5.51
1.79	1.73	1.51	1.20
37	21	41	43
4.76	4.09	3.79	3.73*
1.53	1.65	1.55	1.73
38	23	42	41
4.90	5.26	5.20	5.10
1.39	1.14	1.31	1.12
39	23	42	42
4.77	5.30	5.14	5.15
1.53	1.22	1.36	1.17
39	23	43	40
3.47	4.59	3.55	3.23*
1.58	1.47	1.76	1.66
36	22	42	39
4.67	4.29	5.17	5.40*
1.80	1.45	1.69	1.50
37	21	41	42
2.48	2.05	5.53	2.93*
1.50	1.32	1.54	1.49
33	20	40	42
3.75	3.32	4.58	4.37*
1.78	1.46	1.92	1.50
36	22	40	41
4.24	4.35	5.55	5.90*
1.92	1.35	1.35	1.27
34	20	43	42

Future Emphasis on Programs and Activities by Industry
(U.S.) (*continued*)

	Industry	
Programs / Activities	Total	Consumer
Statistical quality control		
Mean	5.40	5.39
Standard deviation	1.42	1.26
Number	177	31
Closing and/or relocating plants		
Mean	3.05	3.54
Standard deviation	2.13	2.49
Number	159	26
Quality circles		
Mean	3.52	3.50
Standard deviation	1.93	2.14
Number	159	26
Investing in improved production-inventory control systems		
Mean	4.55	4.74
Standard deviation	1.51	1.58
Number	168	27
Hiring in new skills from outside		
Mean	3.78	3.42
Standard deviation	1.64	1.70
Number	166	26
Linking manufacturing strategy to business strategy		
Mean	5.60	5.35
Standard deviation	1.24	1.45
Number	177	31

	Industry		
Industrial Goods	*Basic*	*Machinery*	*Electronics*
5.26	5.96	5.29	5.35
1.88	1.02	1.42	1.23
38	23	42	43
2.71	3.05	2.88	3.21
2.02	2.06	2.06	2.09
34	20	41	38
3.52	3.59	3.67	3.36
2.12	2.15	1.75	1.74
33	22	39	39
4.85	4.48	4.29	4.43
1.27	1.36	1.54	1.72
39	21	41	40
3.84	4.40	3.69	3.73
1.46	1.57	1.51	1.90
37	20	42	41
5.64	5.79	5.63	5.62
1.06	0.98	1.34	1.29
39	24	41	42

*The ANOVA F-test calculated from the comparison between industries within the United States is significant at the .05 alpha level.

(concluded)

Internal Environment by Industry (U.S.)

Indicators	Industry		
	Total	Consumer	Industrial Goods
Top management only			
Mean	0.01	0.00	0.00
Standard deviation	0.07	0.00	0.00
Number	182	32	39
Top and some middle management			
Mean	0.24	0.28	0.31
Standard deviation	0.43	0.46	0.47
Number	182	32	39
Top and most middle management			
Mean	0.43	0.56	0.26
Standard deviation	0.50	0.50	0.44
Number	182	32	39
Every manager and supervisor			
Mean	0.20	0.13	0.21
Standard deviation	0.40	0.34	0.41
Number	182	32	39
Every manager, supervisor, and worker			
Mean	0.13	0.03	0.23
Standard deviation	0.34	0.18	0.43
Number	182	32	39

	Industry	
Basic	*Machinery*	*Electronics*
0.04	0.00	0.00
0.20	0.00	0.00
24	43	44
0.17	0.26	0.16
0.38	0.44	0.37
24	43	44
0.54	0.35	0.50*
0.51	0.48	0.51
24	43	44
0.13	0.21	0.27
0.34	0.41	0.45
24	43	44
0.13	0.19	0.07
0.34	0.39	0.26
24	43	44

*The ANOVA F-test calculated from the comparison between industries is significant at the .05 alpha level.

Past and Future Percentage of Domestic and Foreign Sales, Productions and Purchases (U.S.)

Competitive Abilities	Industry	
	Total	Consumer
Domestic sales, 1989		
Mean	85.41	92.87
Standard deviation	59.94	12.65
Number	160	29
Foreign sales, 1989		
Mean	21.39	7.13
Standard deviation	35.61	12.65
Number	160	29
Domestic sales, 1992		
Mean	82.42	91.12
Standard deviation	69.18	13.44
Number	155	28
Foreign sales, 1992		
Mean	25.90	8.88
Standard deviation	43.43	13.44
Number	155	28
Domestic production, 1989		
Mean	89.30	91.70
Standard deviation	16.12	13.48
Number	155	29
Foreign production, 1989		
Mean	10.70	8.30
Standard deviation	16.12	13.48
Number	155	29
Domestic production, 1992		
Mean	85.94	90.19
Standard deviation	19.21	14.03
Number	152	28
Foreign production, 1992		
Mean	14.06	9.81
Standard deviation	19.21	14.03
Number	152	28
Domestic purchases, 1989		
Mean	84.25	86.64
Standard deviation	18.18	17.91
Number	153	29
Foreign purchases, 1989		
Mean	15.75	13.36
Standard deviation	18.18	17.91
Number	153	29

	Industry		
Industrial Goods	Basic	Machinery	Electronics
85.97	80.05	96.37	72.43
18.81	19.54	123.67	18.93
37	19	35	40
14.03	19.95	35.00	27.33*
18.81	19.54	65.86	19.06
37	19	35	40
82.66	78.31	94.06	66.90
19.81	20.38	139.30	20.74
36	16	36	39
17.34	21.69	42.05	32.85*
19.81	20.38	80.49	20.89
36	16	36	39
90.28	90.88	91.34	83.97
14.54	17.33	13.73	20.01
37	17	34	38
9.72	9.12	8.66	16.03
14.54	17.33	13.73	20.01
37	17	34	38
87.01	89.25	85.76	80.43
18.99	19.34	21.30	20.32
36	16	35	37
12.99	10.75	14.24	19.57
18.99	19.34	21.30	20.32
36	16	35	37
87.69	84.99	84.66	78.32
11.79	18.28	18.18	22.50
36	17	34	37
12.31	15.01	15.34	21.68
11.79	18.28	18.18	22.50
36	17	34	37

Past and Future Percentage of Domestic and Foreign Sales, Productions and Purchases (U.S.) (*continued*)

Competitive Abilities	Industry	
	Total	*Consumer*
Domestic purchases, 1992		
Mean	82.21	85.11
Standard deviation	17.88	17.70
Number	160	29
Foreign purchases, 1992		
Mean	17.79	14.89
Standard deviation	17.88	17.70
Number	160	29

	Industry		
Industrial Goods	**Basic**	**Machinery**	**Electronics**
85.97	85.37	82.13	74.97
14.21	16.23	16.29	21.75
37	19	35	40
14.03	14.63	17.87	25.03
14.21	16.23	16.29	21.75
37	19	35	40

*The ANOVA F-test calculated from the comparison between industries within the United States is significant at the .05 alpha level.

(*concluded*)

Business Unit Growth Strategy by Industry (Europe)

Indicators	Industry		
	Total	Consumer	Industrial Goods
Build market share			
Mean	0.66	0.79	0.67
Standard deviation	0.47	0.41	0.49
Number	224	47	18
Hold (defend) market share			
Mean	0.31	0.19	0.33
Standard deviation	0.46	0.40	0.49
Number	224	47	18
Harvest (maximize cash flow, sacrifice market share)			
Mean	0.02	0.00	0.00
Standard deviation	0.15	0.00	0.00
Number	224	47	18
Withdraw (prepare to exit the business)			
Mean	0.00	0.00	0.00
Standard deviation	0.07	0.00	0.00
Number	224	47	18

Industry		
Basic	Machinery	Electronics
0.50	0.70	0.73
0.50	0.46	0.45
68	43	48
0.46	0.26	0.25
0.50	0.44	0.43
68	43	48
0.04	0.02	0.02
0.21	0.15	0.14
68	43	48
0.00	0.02	0.00
0.00	0.15	0.00
68	43	48

*The ANOVA F-test calculated from the comparison between industries is significant at the .05 alpha level.

Selected Business Unit Profile by Industry (Europe)

	Industry	
Indicators	Total	Consumer
Annual sales revenue[†]		
Mean	15,588.74	12,368.98
Standard deviation	73,789.70	36,151.49
Number	207	42
Pre-tax return on assets (% of assets)		
Mean	18.05	16.59
Standard deviation	25.34	16.69
Number	152	32
Net pre-tax profit ratio (% of sales)		
Mean	8.84	8.00
Standard deviation	10.26	8.08
Number	173	32
R&D expenses (% of sales)		
Mean	4.54	3.03
Standard deviation	3.88	3.77
Number	166	32
Growth rate in unit sales (%)		
Mean	14.91	17.31
Standard deviation	29.99	49.85
Number	196	39
Market share of primary product		
Mean	29.30	32.49
Standard deviation	21.50	21.02
Number	182	37
Capacity utilization (%)		
Mean	84.34	80.35
Standard deviation	22.23	38.37
Number	194	40
Number of plants in the business unit		
Mean	8.44	6.80
Standard deviation	23.00	8.96
Number	210	44
Number of employees		
Mean	6,374.78	2,629.09
Standard deviation	27,798.75	4,107.07
Number	215	44
Manufacturing direct labor employees		
Mean	2,846.73	894.97
Standard deviation	14,322.03	1,201.67
Number	195	39

	Industry		
Industrial Goods	*Basic*	*Machinery*	*Electronics*
309.76	27,202.48	14,312.48	9,631.61
442.12	122,951.75	36,394.81	42,110.52
17	62	40	46
15.50	19.38	13.12	22.66
9.20	21.04	16.54	43.26
14	48	26	32
9.00	12.68	6.45	6.33*
3.82	12.76	6.40	11.18
15	53	33	40
4.07	3.82	3.67	7.68*
3.00	3.16	1.95	4.72
14	49	33	38
7.07	10.38	12.00	24.47
12.85	13.88	9.26	36.93
15	61	38	43
34.60	30.07	29.72	25.57
21.82	24.53	19.81	18.78
15	54	39	37
79.38	83.73	92.23	83.28
16.72	16.41	13.56	13.93
16	59	40	39
3.56	14.73	8.71	2.56
4.93	38.43	14.71	3.09
18	64	41	43
537.28	2,569.97	23,726.05	1,775.76*
506.42	5,353.78	59,821.85	2,558.53
18	65	42	46
368.69	998.02	11,218.87	684.25*
339.45	1,975.94	31,261.31	975.43
16	58	38	44

Selected Business Unit Profile by Industry (Europe) (*continued*)

Indicators	Industry	
	Total	Consumer
Manufacturing indirect labor employees		
Mean	656.75	379.69
Standard deviation	1,388.66	752.89
Number	194	39

	Industry		
Industrial Goods	*Basic*	*Machinery*	*Electronics*
117.31	632.81	1,403.75	519.39*
173.79	1,344.81	2,302.12	851.75
16	59	36	44

†In U.S. dollars.
*The ANOVA F-test calculated from the comparison between industries within Europe is significant at the .05 alpha level.

(*concluded*)

Importance of Competitive Abilities by Industry (Europe)

Importance of Competitive Abilities	Industry	
	Total	Consumer
Ability to profit in price competitive market		
Mean	5.51	5.20
Standard deviation	1.30	1.75
Number	221	46
Ability to make rapid changes in design		
Mean	4.88	4.66
Standard deviation	1.52	1.66
Number	216	44
Ability to introduce new products quickly		
Mean	5.31	5.49
Standard deviation	1.48	1.44
Number	220	47
Ability to make rapid volume changes		
Mean	4.96	5.00
Standard deviation	1.53	1.55
Number	220	47
Ability to make rapid product mix changes		
Mean	4.92	4.89
Standard deviation	1.44	1.45
Number	219	47
Ability to offer a broad product line		
Mean	5.05	4.94
Standard deviation	1.49	1.57
Number	222	47
Ability to offer consistently low defect rates		
Mean	6.36	6.28
Standard deviation	1.00	1.33
Number	222	47
Ability to provide high-performance products or product amenities		
Mean	5.68	5.61
Standard deviation	1.28	1.57
Number	218	44
Ability to provide reliable/durable products		
Mean	6.00	5.95
Standard deviation	1.24	1.38
Number	214	42
Ability to provide fast deliveries		
Mean	5.64	5.57
Standard deviation	1.13	1.36
Number	222	47

| Industry | | | |
Industrial Goods	Basic	Machinery	Electronics
5.72	5.47	5.56	5.73
1.32	1.15	1.05	1.16
18	68	41	48
4.61	4.24	5.61	5.43*
1.42	1.62	1.02	1.19
18	66	41	47
5.06	4.86	5.55	5.64*
1.35	1.62	1.25	1.42
18	66	42	47
5.24	4.62	5.12	5.19
1.56	0.55	1.27	1.68
17	68	41	47
4.39	4.69	5.17	5.25
1.24	1.44	1.18	1.62
18	65	41	48
4.89	4.88	5.19	5.31
1.60	1.46	1.55	1.36
18	67	42	48
6.39	6.22	6.60	6.40
0.85	0.92	0.63	1.05
18	67	42	48
5.56	5.45	6.07	5.77
1.04	1.27	1.00	1.24
18	66	42	48
5.89	5.65	6.29	6.29*
1.23	1.37	0.98	0.99
18	65	41	48
5.28	5.75	5.74	5.60
1.02	1.13	1.06	0.98
18	67	42	48

Importance of Competitive Abilities by Industry (Europe) (*continued*)

	Industry	
Importance of Competitive Abilities	*Total*	*Consumer*
Ability to make dependable delivery promises		
Mean	6.21	6.07
Standard deviation	1.01	1.29
Number	221	46
Ability to provide effective after-sales service		
Mean	5.25	4.26
Standard deviation	1.62	1.72
Number	213	43
Ability to provide product support effectively		
Mean	5.24	4.81
Standard deviation	1.33	1.40
Number	211	42
Ability to make product easily available (broad distribution)		
Mean	4.93	5.11
Standard deviation	1.58	1.65
Number	216	46
Ability to customize products and services to customer needs		
Mean	5.57	5.07
Standard deviation	1.39	1.67
Number	217	46

	Industry		
Industrial Goods	*Basic*	*Machinery*	*Electronics*
6.44	6.12	6.31	6.29
0.51	1.14	0.68	0.87
18	67	42	48
4.11	5.05	6.24	6.04*
1.49	1.53	1.11	1.11
18	65	41	46
4.88	5.20	5.78	5.74*
1.27	1.28	1.28	1.20
17	65	41	46
4.76	4.92	4.90	4.83
1.86	1.38	1.55	1.74
17	66	41	46
5.17	5.44	6.00	6.02*
1.58	1.37	0.87	1.18
18	66	41	46

*The ANOVA F-test calculated from the comparison between industries within Europe is significant at the .05 alpha level.

(concluded)

Strength in Competitive Abilities by Industry (Europe)

Competitive Abilities	Industry	
	Total	Consumer
Ability to profit in price-competitive market		
Mean	4.37	4.49
Standard deviation	1.29	1.38
Number	219	45
Ability to make rapid changes in design		
Mean	4.48	4.16
Standard deviation	1.31	1.28
Number	218	45
Ability to introduce new products quickly		
Mean	4.30	4.13
Standard deviation	1.31	1.24
Number	220	47
Ability to make rapid volume changes		
Mean	4.57	4.54
Standard deviation	1.25	1.17
Number	216	46
Ability to make rapid product mix changes		
Mean	4.64	4.72
Standard deviation	1.24	1.10
Number	218	47
Ability to offer a broad product line		
Mean	4.83	4.77
Standard deviation	1.55	1.71
Number	220	47
Ability to offer consistently low defect rates		
Mean	5.26	5.26
Standard deviation	1.20	1.34
Number	221	46
Ability to provide high-performance products or product amenities		
Mean	5.24	5.05
Standard deviation	1.14	1.12
Number	218	44
Ability to provide reliable/durable products		
Mean	5.40	5.26
Standard deviation	1.22	1.42
Number	216	43
Ability to provide fast deliveries		
Mean	4.75	4.81
Standard deviation	1.28	1.21
Number	221	47

	Industry		
Industrial Goods	*Basic*	*Machinery*	*Electronics*
4.00	4.64	4.05	4.29
1.03	1.21	1.12	1.47
18	67	41	48
4.67	4.52	4.57	4.57
1.08	1.41	1.29	1.31
18	66	42	47
4.22	4.47	4.36	4.21
1.22	1.33	1.36	1.37
18	66	42	47
4.76	4.46	4.78	4.49
1.03	1.34	1.15	1.36
17	67	41	45
4.61	4.66	4.54	4.64
0.98	1.24	1.40	1.37
18	65	41	47
4.88	4.82	4.86	4.87
1.36	1.46	1.62	1.60
17	67	42	47
5.56	5.12	5.14	5.44
1.25	1.14	0.90	1.34
18	67	42	48
5.06	5.06	5.64	5.40*
1.16	1.16	0.93	1.20
18	66	42	48
5.17	5.25	5.60	5.65
1.04	1.17	0.91	1.34
18	65	42	48
4.72	4.92	4.69	4.52
1.32	1.24	1.41	1.27
18	66	42	48

Strength in Competitive Abilities by Industry (Europe) (*continued*)

	Industry	
Competitive Abilities	*Total*	*Consumer*
Ability to make dependable delivery promises		
Mean	4.90	4.85
Standard deviation	1.34	1.35
Number	221	46
Ability to provide effective after-sales service		
Mean	4.90	4.55
Standard deviation	1.32	1.19
Number	210	42
Ability to provide product support effectively		
Mean	4.88	4.68
Standard deviation	1.23	1.27
Number	209	40
Ability to make products easily available (broad distribution)		
Mean	4.72	5.13
Standard deviation	1.31	1.25
Number	215	45
Ability to customize products and services to customer needs		
Mean	4.86	4.47
Standard deviation	1.33	1.16
Number	214	45

Industry			
Industrial Goods	*Basic*	*Machinery*	*Electronics*
5.06	5.01	5.12	4.54
1.16	1.31	1.23	1.49
18	67	42	48
5.00	4.72	5.15	5.24
1.33	1.45	1.31	1.17
18	64	41	45
4.50	4.92	5.05	5.00
1.62	1.19	1.16	1.14
18	64	41	46
4.53	4.71	4.55	4.54
1.37	1.22	1.43	1.33
17	65	42	46
4.83	4.71	5.38	5.02*
1.29	1.52	1.08	1.31
18	65	40	46

*The ANOVA F-test calculated from the comparison between industries within Europe is significant at the .05 alpha level.

(concluded)

Manufacturing Cost Structure by Industry (Europe)

	Industry	
Manufacturing Cost Structure	Total	Consumer
Manufacturing costs as percent of sales, 1989		
Mean	51.76	44.51
Standard deviation	22.04	20.38
Number	190	41
Manufacturing costs as percent of sales, 1992		
Mean	49.78	41.13
Standard deviation	22.15	20.49
Number	184	39
Materials cost (% of total mfg. cost), 1989		
Mean	54.82	58.71
Standard deviation	21.21	22.60
Number	195	41
Materials cost (% of total mfg. cost), 1992		
Mean	57.46	57.64
Standard deviation	21.10	22.98
Number	186	39
Direct labor cost (% of total mfg. cost), 1989		
Mean	18.05	17.66
Standard deviation	12.66	12.47
Number	196	41
Direct labor cost (% of total mfg. cost), 1992		
Mean	16.76	17.03
Standard deviation	12.17	11.98
Number	186	38
Energy cost (% of total mfg. cost), 1989		
Mean	4.81	3.50
Standard deviation	6.83	3.65
Number	191	40
Energy cost (% of total mfg. cost), 1992		
Mean	4.81	3.63
Standard deviation	7.00	3.39
Number	185	38
Manufacturing overhead cost (% of total mfg. cost), 1989		
Mean	21.36	20.45
Standard deviation	14.17	14.24
Number	192	40

	Industry		
Industrial Goods	Basic	Machinery	Electronics
50.36	54.95	58.03	49.28
23.69	24.32	22.89	16.67
14	58	37	40
49.75	53.47	56.75	46.51*
25.43	24.20	23.00	15.36
12	58	36	39
43.63	57.50	49.14	56.39*
11.67	21.51	21.64	20.32
16	58	36	44
47.69	59.18	52.57	62.00
11.68	21.85	19.74	20.69
13	56	35	43
22.88	15.93	22.83	15.59*
13.26	10.36	14.84	12.51
16	59	36	44
23.69	14.39	21.77	13.49*
13.29	9.71	14.63	11.03
13	57	35	43
3.81	8.62	3.03	2.63*
3.39	10.35	2.57	3.87
16	58	36	41
3.50	8.64	3.11	2.60*
3.32	10.79	2.83	3.70
14	56	35	42
27.00	17.60	23.03	23.93
10.99	12.08	17.19	14.12
15	58	36	43

Manufacturing Cost Structure by Industry (Europe) (*continued*)

	Industry	
Manufacturing Cost Structure	*Total*	*Consumer*
Manufacturing overhead cost (% of total mfg. cost), 1992		
Mean	21.23	20.76
Standard deviation	14.13	14.53
Number	190	38

		Industry	
Industrial Goods	*Basic*	*Machinery*	*Electronics*
27.93	17.62	22.64	23.27
11.06	11.59	16.45	15.08
15	60	36	41

*The ANOVA F-test calculated from the comparison between industries within Europe is significant at the .05 alpha level.

(*concluded*)

Manufacturing Performance Improvement Index by Industry (Europe)

Performance Indicators	Industry	
	Total	Consumer
Overall quality as perceived by customers		
Mean	113.06	109.69
Standard deviation	16.15	12.95
Number	216	45
Average unit production costs for typical product		
Mean	108.05	111.88
Standard deviation	18.98	21.26
Number	213	43
Inventory turnover		
Mean	114.75	11.84
Standard deviation	23.93	22.94
Number	211	44
Speed of new product development and/or design change		
Mean	107.61	110.29
Standard deviation	19.02	23.44
Number	209	42
On-time delivery		
Mean	114.55	110.25
Standard deviation	39.09	12.09
Number	212	44
Equipment changeover (or set-up) time		
Mean	110.98	105.37
Standard deviation	22.14	19.92
Number	203	41
Market share		
Mean	110.80	111.95
Standard deviation	19.26	17.25
Number	210	43
Profitability		
Mean	128.06	126.98
Standard deviation	65.43	44.33
Number	199	41
Customer service		
Mean	114.48	113.24
Standard deviation	17.38	18.06
Number	204	42
Manufacturing lead time		
Mean	113.20	111.57
Standard deviation	19.93	14.79
Number	211	44

	Industry		
Industrial Goods	Basic	Machinery	Electronics
111.11	110.89	116.24	117.26
8.84	9.98	18.52	23.49
18	65	42	46
105.06	102.72	109.69	111.61
11.40	11.89	27.90	15.64
17	65	42	46
116.94	109.84	119.50	119.00
24.09	16.28	28.23	28.21
17	62	42	46
108.50	105.09	107.80	108.18
13.57	14.44	26.01	14.19
18	64	41	44
110.56	109.88	123.71	118.67
14.03	14.18	81.20	26.00
18	64	41	45
107.50	108.21	118.38	114.84*
12.63	12.90	34.39	21.54
18	61	40	43
106.39	104.97	112.19	118.50*
14.65	11.03	21.58	26.20
18	63	42	44
116.61	135.05	124.77	126.67
38.36	97.21	51.19	40.43
18	62	39	39
110.56	111.62	117.22	119.25
10.42	11.88	22.43	19.69
18	63	41	40
102.06	109.90	113.57	123.00*
14.37	11.38	13.78	32.55
17	62	42	46

Manufacturing Performance Improvement Index by Industry (Europe) (*continued*)

	Industry	
Performance Indicators	*Total*	*Consumer*
Procurement lead time		
Mean	107.61	106.19
Standard deviation	15.81	10.52
Number	209	42
Delivery lead time		
Mean	110.13	105.30
Standard deviation	18.51	12.74
Number	211	44
Variety of products producible by manufacturing		
Mean	112.79	114.00
Standard deviation	25.06	21.24
Number	199	41

	Industry		
Industrial Goods	*Basic*	*Machinery*	*Electronics*
105.71	105.59	107.07	113.00
13.17	10.57	11.61	26.32
17	63	42	45
105.11	109.70	109.66	118.07*
15.40	16.45	18.68	24.44
18	64	41	44
106.76	111.69	113.08	115.55
18.54	26.60	25.94	28.21
17	62	39	40

*The ANOVA F-test calculated from the comparison between industries within Europe is significant at the .05 alpha level.

(*concluded*)

Manufacturing Objectives by Industry (Europe)

	Industry	
Manufacturing's Objectives	Total	Consumer
Improve conformance quality (reduce defects)		
Mean	5.83	5.78
Standard deviation	1.10	1.01
Number	222	46
Reduce unit costs		
Mean	5.70	5.40
Standard deviation	1.15	1.41
Number	223	47
Improve safety record		
Mean	4.49	4.58
Standard deviation	1.60	1.42
Number	220	45
Reduce manufacturing lead time		
Mean	5.08	5.15
Standard deviation	1.48	1.33
Number	221	46
Increase capacity		
Mean	4.37	4.64
Standard deviation	1.66	1.65
Number	222	47
Reduce procurement lead time		
Mean	4.76	4.79
Standard deviation	1.46	1.33
Number	220	47
Reduce new product development cycle		
Mean	5.20	5.26
Standard deviation	1.43	1.63
Number	221	46
Reduce materials costs		
Mean	5.19	4.70
Standard deviation	1.33	1.49
Number	220	47
Reduce overhead costs		
Mean	5.56	5.74
Standard deviation	1.25	1.18
Number	222	46
Improve direct labor productivity		
Mean	5.45	5.61
Standard deviation	1.19	1.20
Number	220	46

	Industry		
Industrial Goods	*Basic*	*Machinery*	*Electronics*
5.67	5.75	6.14	5.77
1.08	1.26	0.80	1.18
18	68	43	47
6.11	5.51	5.91	5.91*
1.02	1.11	0.92	1.06
18	68	43	47
4.44	4.81	4.37	4.07
1.65	1.75	1.43	1.62
18	68	43	46
4.78	4.46	5.49	5.64*
1.90	1.68	1.10	1.11
18	67	43	47
3.72	4.63	4.51	3.85*
1.87	1.66	1.71	1.43
18	68	43	46
4.44	4.18	5.14	5.38*
1.42	1.64	1.19	1.19
18	67	43	45
4.83	4.78	5.51	5.62*
1.50	1.35	1.08	1.45
18	67	43	47
4.94	5.20	5.33	5.64*
1.26	1.28	1.34	1.13
18	65	43	47
5.33	5.43	5.56	5.66
1.68	1.18	1.35	1.15
18	68	43	47
5.67	5.35	5.58	5.21
1.19	1.12	1.14	1.32
18	66	43	47

Manufacturing Objectives by Industry (Europe) (*continued*)

	Industry	
Manufacturing's Objectives	*Total*	*Consumer*
Increase throughput		
Mean	5.34	5.47
Standard deviation	1.26	1.16
Number	220	47
Reduce number of vendors		
Mean	3.86	4.06
Standard deviation	1.52	1.29
Number	220	47
Improve vendor quality		
Mean	5.35	5.66
Standard deviation	1.27	1.22
Number	220	47
Reduce inventories		
Mean	5.01	5.06
Standard deviation	1.53	1.45
Number	221	47
Increase delivery reliability		
Mean	5.54	5.49
Standard deviation	1.36	1.44
Number	223	47
Increase delivery speed		
Mean	5.05	4.91
Standard deviation	1.31	1.38
Number	223	47
Improve ability to make rapid product mix changes		
Mean	4.79	4.87
Standard deviation	1.44	1.57
Number	222	46
Improve ability to make rapid volume changes		
Mean	4.60	4.76
Standard deviation	1.37	1.34
Number	221	46
Reduce break-even points		
Mean	4.72	4.64
Standard deviation	1.37	1.38
Number	218	45
Raise employee morale		
Mean	5.04	5.09
Standard deviation	1.18	1.12
Number	223	47

	Industry		
Industrial Goods	*Basic*	*Machinery*	*Electronics*
5.33	5.01	5.74	5.33
1.37	1.43	0.96	1.19
18	67	42	46
3.82	3.27	4.02	4.34*
1.47	1.43	1.67	1.52
17	66	43	47
4.88	4.86	5.56	5.68*
1.17	1.41	1.03	1.14
17	66	43	47
5.06	4.51	5.21	5.47*
1.68	1.70	1.39	1.28
17	67	43	47
5.28	5.29	5.67	5.94
1.74	1.44	1.17	1.09
18	68	43	47
4.83	4.69	5.42	5.45*
1.29	1.35	1.22	1.12
18	68	43	47
4.72	4.41	4.93	5.17
1.45	1.58	1.14	1.27
18	68	43	47
4.44	4.33	4.77	4.74
1.34	1.49	1.25	1.31
18	67	43	47
4.17	4.62	5.12	4.81
1.54	1.42	1.33	1.21
18	65	43	47
4.89	4.84	5.42	4.98
1.02	1.23	1.07	1.28
18	68	43	47

Manufacturing Objectives by Industry (Europe) (*continued*)

	Industry	
Manufacturing's Objectives	*Total*	*Consumer*
Maximize cash flow		
Mean	5.13	5.11
Standard deviation	1.33	1.29
Number	219	46
Increase environmental safety and protection		
Mean	4.84	4.89
Standard deviation	1.50	1.36
Number	223	47
Reduce capacity		
Mean	2.23	2.19
Standard deviation	1.45	1.48
Number	221	47
Increase product or materials standardization		
Mean	4.64	4.70
Standard deviation	1.45	1.44
Number	222	47
Improve labor relations		
Mean	4.44	4.38
Standard deviation	1.42	1.51
Number	222	47
Improve white-collar productivity		
Mean	5.23	5.00
Standard deviation	1.17	1.40
Number	220	47
Increase range of products produced by existing facilities		
Mean	3.91	3.87
Standard deviation	1.64	1.58
Number	222	47
Meet financial shipping goals		
Mean	4.55	4.48
Standard deviation	1.51	1.49
Number	210	42
Improve pre-sales service and technical support		
Mean	4.62	4.02
Standard deviation	1.43	1.61
Number	214	43
Improve after-sales service		
Mean	4.74	4.05
Standard deviation	1.48	1.60
Number	215	43

	Industry		
Industrial Goods	*Basic*	*Machinery*	*Electronics*
5.22	4.88	5.52	5.13
1.66	1.39	1.04	1.32
18	68	42	45
5.28	5.46	4.56	3.98*
1.18	1.41	1.33	1.58
18	68	43	47
2.28	1.94	2.17	2.70
1.49	1.49	1.29	1.44
18	67	42	47
4.56	4.10	5.07	4.98*
1.50	1.56	1.16	1.31
18	67	43	47
4.39	4.28	4.91	4.32
1.29	1.35	1.36	1.49
18	67	43	47
5.00	5.09	5.62	5.40
1.32	1.06	0.99	1.12
17	67	42	47
3.72	4.07	3.77	3.89
1.78	1.72	1.72	1.48
18	67	43	47
4.31	4.19	4.59	5.20*
1.54	1.50	1.43	1.45
16	67	39	46
4.17	4.48	5.19	5.02*
1.69	1.34	1.03	1.36
18	66	43	44
3.89	4.67	5.47	5.16*
1.75	1.34	1.10	1.35
18	67	43	44

Manufacturing Objectives by Industry (Europe) (*continued*)

Manufacturing's Objectives	Industry	
	Total	Consumer
Change culture of manufacturing organization		
Mean	5.08	5.17
Standard deviation	1.54	1.67
Number	222	47
Improve interfunctional communication		
Mean	5.49	5.53
Standard deviation	1.11	1.20
Number	223	47
Improve communication with external partners		
Mean	5.08	4.91
Standard deviation	1.30	1.23
Number	223	47
Reduce set-up/changeover times		
Mean	4.88	4.85
Standard deviation	1.57	1.78
Number	220	47

	Industry		
Industrial Goods	**Basic**	**Machinery**	**Electronics**
5.39	4.99	5.12	4.98
1.69	1.54	1.48	1.42
18	67	43	47
5.17	5.37	5.56	5.70
1.34	1.04	0.98	1.12
18	68	43	47
4.89	4.90	5.35	5.34
1.18	1.54	1.09	1.18
18	68	43	47
5.06	4.70	5.10	4.89
1.85	1.45	1.43	1.54
17	67	42	47

*The ANOVA F-test calculated from the comparison between industries within Europe is significant at the .05 alpha level.

(*concluded*)

Past Pay-offs from Programs and Activities by Industry (Europe)

Programs / Activities	Industry	
	Total	Consumer
Giving workers a broad range of tasks and/or more responsibility		
Mean	4.06	3.97
Standard deviation	1.46	1.50
Number	172	35
Activity-based costing		
Mean	3.63	3.73
Standard deviation	1.61	1.58
Number	130	22
Manufacturing reorganization		
Mean	4.58	4.72
Standard deviation	1.44	1.34
Number	179	36
Worker training		
Mean	4.52	4.44
Standard deviation	1.30	1.32
Number	183	36
Management training		
Mean	4.62	4.68
Standard deviation	1.26	1.16
Number	183	37
Supervisor training		
Mean	4.66	4.71
Standard deviation	1.23	0.99
Number	175	35
Computer-aided manufacturing (CAM)		
Mean	3.96	3.66
Standard deviation	1.69	1.67
Number	164	29
Computer-aided design (CAD)		
Mean	3.90	3.74
Standard deviation	1.79	1.68
Number	156	23
Value analysis/product redesign		
Mean	3.75	3.48
Standard deviation	1.56	1.62
Number	138	23
Interfunctional work teams		
Mean	4.27	4.73
Standard deviation	1.43	1.15
Number	154	26

	Industry		
Industrial Goods	Basic	Machinery	Electronics
4.64	3.89	4.33	3.94
1.34	1.47	1.24	1.63
14	55	33	35
3.56	3.54	4.30	3.20
2.19	1.52	1.52	1.61
9	46	23	30
4.50	4.30	5.06	4.42
1.65	1.52	1.16	1.48
14	56	35	38
4.44	4.49	4.60	4.64
1.21	1.40	1.26	1.22
16	63	35	33
4.29	4.55	4.90	4.52
1.90	1.19	1.21	1.28
14	60	39	33
4.50	4.66	4.73	4.64
1.37	1.21	1.31	1.41
16	58	33	33
2.83	3.95	4.73	3.91*
1.80	1.59	1.64	1.66
12	56	33	34
1.78	3.12	5.03	4.43*
1.64	1.64	1.40	1.52
9	50	37	37
2.56	3.54	4.52	3.83*
1.59	1.52	1.21	1.61
9	46	31	29
3.58	4.24	4.77	3.70*
1.56	1.40	1.15	1.66
12	55	31	30

**Past Pay-offs from Programs and Activities by Industry
(Europe)** (*continued*)

Programs / Activities	Industry	
	Total	Consumer
Quality function deployment		
Mean	4.52	4.41
Standard deviation	1.43	1.50
Number	174	32
Developing new processes for new products		
Mean	4.43	4.63
Standard deviation	1.58	1.40
Number	164	30
Developing new processes for old products		
Mean	4.16	4.56
Standard deviation	1.66	1.44
Number	158	36
Integrating information systems in manufacturing		
Mean	4.41	4.24
Standard deviation	1.41	1.56
Number	174	33
Integrating information systems across functions		
Mean	4.23	4.24
Standard deviation	1.50	1.38
Number	165	29
Reconditioning physical plants		
Mean	4.29	4.53
Standard deviation	1.59	1.52
Number	153	34
Just-in-time		
Mean	4.21	4.24
Standard deviation	1.71	1.75
Number	154	29
Robots		
Mean	3.04	2.45
Standard deviation	1.98	1.82
Number	125	20
Flexible manufacturing systems		
Mean	3.90	3.87
Standard deviation	1.80	1.55
Number	144	23
Design for manufacture		
Mean	3.52	2.85
Standard deviation	1.70	1.60
Number	118	20

	Industry		
Industrial Goods	*Basic*	*Machinery*	*Electronics*
3.64	4.42	5.09	4.53*
1.29	1.45	1.22	1.43
11	59	34	38
3.82	4.37	4.33	4.61
1.17	1.72	1.41	1.76
11	54	33	36
4.09	4.07	4.30	3.65
1.58	1.79	1.51	1.79
11	55	30	26
3.36	4.32	4.49	4.92*
1.12	1.35	1.48	1.21
11	56	37	37
3.00	4.43	4.11	4.46*
1.28	1.43	1.55	1.60
12	53	36	35
3.75	4.25	4.58	3.94
1.58	1.63	1.34	1.81
8	48	31	32
3.92	3.67	4.34	4.86*
1.68	1.67	1.58	1.73
12	45	32	36
3.14	2.56	4.14	3.04*
2.12	1.94	1.77	1.97
7	43	29	26
3.33	3.45	4.41	4.19
1.66	1.82	1.46	2.17
9	47	34	31
3.40	2.71	4.48	4.18*
1.17	1.41	1.73	1.61
10	35	25	28

Past Pay-offs from Programs and Activities by Industry (Europe) (*continued*)

Programs / Activities	Industry	
	Total	Consumer
Statistical quality control		
Mean	4.25	4.34
Standard deviation	1.52	1.23
Number	162	29
Closing and/or relocating plants		
Mean	3.07	3.05
Standard deviation	2.05	1.82
Number	121	20
Quality circles		
Mean	3.67	4.00
Standard deviation	1.82	1.74
Number	146	26
Investing in improved production-inventory control systems		
Mean	4.09	4.25
Standard deviation	1.65	1.59
Number	147	24
Hiring in new skills from outside		
Mean	3.78	4.28
Standard deviation	1.72	1.67
Number	147	29
Linking manufacturing strategy to business strategy		
Mean	4.57	4.69
Standard deviation	1.54	1.49
Number	169	32

	Industry		
Industrial Goods	*Basic*	*Machinery*	*Electronics*
4.08	3.82	4.48	4.72
1.88	1.54	1.50	1.46
12	56	33	32
4.11	3.15	3.27	2.36
2.52	2.21	1.95	1.78
9	41	26	25
2.75	3.46	4.27	3.41
1.67	1.66	1.80	2.05
8	50	30	32
3.25	4.10	4.23	4.03
1.83	1.56	1.78	1.67
8	52	31	32
3.38	3.28	4.33	3.55*
1.30	1.53	1.74	1.89
8	46	33	31
3.90	4.71	4.38	4.62
1.29	1.61	1.72	1.36
10	56	34	37

*The ANOVA F-test calculated from the comparison between industries within Europe is significant at the .05 alpha level.

(*concluded*)

Future Emphasis on Programs and Activities by Industry (Europe)

| | Industry | |
Programs / Activities	Total	Consumer
Giving workers a broad range of tasks and/or more responsibility		
Mean	4.80	4.93
Standard deviation	1.70	1.75
Number	203	40
Activity-based costing		
Mean	4.26	4.22
Standard deviation	1.77	1.70
Number	179	37
Manufacturing reorganization		
Mean	4.77	4.90
Standard deviation	1.64	1.69
Number	202	40
Worker training		
Mean	5.24	5.29
Standard deviation	1.17	1.31
Number	205	42
Management training		
Mean	5.17	5.16
Standard deviation	1.29	1.59
Number	207	43
Supervisor training		
Mean	5.32	5.50
Standard deviation	1.19	1.23
Number	203	42
Computer-aided manufacturing (CAM)		
Mean	4.54	4.58
Standard deviation	1.92	1.79
Number	193	36
Computer-aided design (CAD)		
Mean	4.37	4.06
Standard deviation	1.93	1.92
Number	196	34
Value analysis/product redesign		
Mean	4.26	4.06
Standard deviation	1.72	1.86
Number	182	35
Interfunctional work teams		
Mean	4.96	5.28
Standard deviation	1.51	1.41
Number	191	36

	Industry		
Industrial Goods	*Basic*	*Machinery*	*Electronics*
4.83	4.57	4.84	5.00
1.86	0.58	1.67	1.82
18	65	38	42
3.67	3.93	4.50	4.79
1.84	1.71	1.85	1.75
15	56	32	39
4.81	4.34	5.03	5.04
1.52	1.73	1.54	1.52
16	65	36	45
5.33	5.35	5.23	4.98
0.91	1.08	0.96	1.42
18	65	39	41
4.76	5.20	5.23	5.24
1.64	1.15	1.05	1.25
17	65	40	42
5.00	5.30	5.31	5.29
1.37	1.14	0.98	1.33
18	66	36	41
3.59	4.55	5.00	4.43
1.84	1.93	1.68	2.19
17	60	40	40
4.00	3.52	5.40	4.98*
1.51	1.84	1.53	1.97
15	62	43	42
3.57	3.75	4.94	4.85*
1.55	1.60	1.45	1.73
14	59	35	39
4.88	4.68	5.15	4.98
1.41	1.49	1.35	1.75
17	63	34	41

**Future Emphasis on Programs and Activities by Industry
(Europe)** (*continued*)

	Industry	
Programs / Activities	*Total*	*Consumer*
Quality function deployment		
Mean	5.30	5.24
Standard deviation	1.55	1.53
Number	197	38
Developing new processes for new products		
Mean	4.98	5.03
Standard deviation	1.70	1.67
Number	203	38
Developing new processes for old products		
Mean	4.31	4.70
Standard deviation	1.81	1.60
Number	190	40
Integrating information systems in manufacturing		
Mean	5.29	5.70
Standard deviation	1.38	1.18
Number	204	40
Integrating information systems across functions		
Mean	5.14	5.26
Standard deviation	1.44	1.29
Number	196	39
Reconditioning physical plants		
Mean	4.40	4.55
Standard deviation	1.67	1.75
Number	188	40
Just-in-time		
Mean	4.88	5.14
Standard deviation	1.69	1.53
Number	193	37
Robots		
Mean	3.12	2.76
Standard deviation	2.06	2.06
Number	175	33
Flexible manufacturing systems		
Mean	4.45	4.51
Standard deviation	1.98	1.77
Number	186	35
Design for manufacture		
Mean	4.11	3.67
Standard deviation	2.06	2.02
Number	168	30

	Industry		
Industrial Goods	*Basic*	*Machinery*	*Electronics*
4.88	5.34	5.49	5.30
1.50	1.62	1.33	1.68
16	62	37	44
4.18	4.88	5.03	5.34
1.88	1.85	1.46	1.60
17	65	39	44
4.56	4.71	4.22	3.19*
1.76	1.69	1.76	1.90
18	59	36	37
4.22	5.22	5.37	5.38*
1.52	1.29	1.04	1.71
18	63	41	42
4.24	5.05	5.41	5.32
1.48	1.41	1.09	1.80
17	63	39	38
4.27	4.36	4.86	3.92
1.44	1.56	1.33	1.99
15	58	36	39
4.61	4.21	5.53	5.02*
1.79	1.69	1.06	1.98
18	56	40	42
2.75	2.61	4.00	3.50*
2.24	1.73	2.04	2.21
16	54	36	36
3.50	3.82	5.32	4.90*
2.09	1.95	1.55	2.10
18	56	37	40
3.56	3.00	5.09	5.23*
1.59	1.65	1.91	1.99
16	49	34	39

**Future Emphasis on Programs and Activities by Industry
(Europe)** (*continued*)

Programs / Activities	Industry	
	Total	Consumer
Statistical quality control		
Mean	4.87	5.03
Standard deviation	1.67	1.40
Number	198	38
Closing and/or relocating plants		
Mean	2.76	2.79
Standard deviation	2.04	2.03
Number	170	33
Quality circles		
Mean	4.23	4.39
Standard deviation	2.07	2.00
Number	180	33
Investing in improved production-inventory control systems		
Mean	4.66	4.97
Standard deviation	1.82	1.85
Number	185	34
Hiring in new skills from outside		
Mean	3.93	4.00
Standard deviation	1.90	1.74
Number	190	38
Linking manufacturing strategy to business strategy		
Mean	5.59	5.68
Standard deviation	1.28	1.17
Number	201	41

	Industry		
Industrial Goods	*Basic*	*Machinery*	*Electronics*
4.94	4.79	5.00	4.70
2.13	1.55	1.49	2.01
18	61	38	43
2.88	3.04	2.75	2.26
2.36	2.06	1.92	2.01
16	54	32	35
3.06	4.05	4.74	4.42
2.41	2.05	1.75	2.14
17	57	35	38
3.71	4.38	4.69	5.23*
2.28	1.77	1.53	1.68
17	60	35	39
3.24	3.38	4.54	4.45*
1.79	1.97	1.76	1.87
17	60	37	38
5.44	5.60	5.32	5.81
1.20	1.49	1.34	1.02
18	62	38	42

*The ANOVA F-test calculated from the comparison between industries within Europe is significant at the .05 alpha level.

(concluded)

Internal Environment by Country (Europe)

Indicators	Industry		
	Total	Consumer	Industrial Goods
Top management only			
Mean	0.07	0.11	0.17
Standard deviation	0.26	0.31	0.38
Number	224	47	18
Top and some middle management			
Mean	0.38	0.36	0.44
Standard deviation	0.49	0.49	0.51
Number	224	47	18
Top and most middle management			
Mean	0.20	0.15	0.11
Standard deviation	0.40	0.36	0.32
Number	224	47	18
Every manager and supervisor			
Mean	0.42	0.45	0.38
Standard deviation	0.42	0.45	0.38
Number	224	47	18
Every manager, supervisor, and worker			
Mean	0.11	0.11	0.06
Standard deviation	0.31	0.31	0.24
Number	224	47	18

	Industry	
Basic	Machinery	Electronics
0.04	0.07	0.04
0.21	0.26	0.20
68	43	48
0.34	0.40	0.40
0.48	0.49	0.49
68	43	48
0.21	0.16	0.29
0.41	0.37	0.46
68	43	48
0.45	0.41	0.38
0.45	0.41	0.38
68	43	48
0.09	0.16	0.10
0.29	0.37	0.31
68	43	48

Past and Future Percentage of Domestic and Foreign Sales, Productions, and Purchases (Europe)

	Industry	
Competitive Abilities	Total	Consumer
Domestic sales, 1989		
Mean	82.59	83.44
Standard deviation	19.93	20.28
Number	182	36
Foreign sales, 1989		
Mean	18.51	16.39
Standard deviation	21.15	20.16
Number	182	36
Domestic sales, 1992		
Mean	80.39	81.94
Standard deviation	20.77	21.40
Number	172	35
Foreign sales, 1992		
Mean	19.95	18.06
Standard deviation	21.04	21.37
Number	172	35
Domestic production, 1989		
Mean	94.00	93.13
Standard deviation	13.50	12.55
Number	184	39
Foreign production, 1989		
Mean	6.17	6.72
Standard deviation	14.18	12.10
Number	183	39
Domestic production, 1992		
Mean	93.07	93.05
Standard deviation	14.05	11.99
Number	177	38
Foreign production, 1992		
Mean	6.86	6.74
Standard deviation	14.03	11.64
Number	176	38
Domestic purchases, 1989		
Mean	82.53	81.11
Standard deviation	23.16	28.15
Number	176	36
Foreign purchases, 1989		
Mean	17.42	18.89
Standard deviation	23.21	28.15
Number	175	36

	Industry		
Industrial Goods	*Basic*	*Machinery*	*Electronics*
92.71	82.05	76.81	84.54
11.50	19.92	22.57	18.21
14	58	37	37
7.29	21.67	23.27	15.08
11.50	23.47	22.60	17.85
14	58	37	37
91.21	79.89	74.31	81.56
13.91	21.09	22.61	18.69
14	53	36	34
8.79	21.34	25.64	18.32
13.91	22.11	22.45	18.58
14	53	36	34
98.82	92.27	93.83	85.49
3.76	16.55	11.01	14.21
17	56	35	37
1.18	8.63	6.17	4.11
3.76	18.47	10.97	14.16
17	56	35	36
94.71	92.06	92.91	93.94
14.94	16.09	12.32	14.65
17	52	34	36
5.29	7.92	7.09	5.97
14.94	16.09	12.32	14.86
17	52	34	35
89.69	82.19	91.63	73.03*
12.72	23.82	9.36	25.99
16	53	34	37
10.31	17.60	8.53	26.97*
12.72	24.02	9.39	25.99
16	52	34	37

Past and Future Percentage of Domestic and Foreign Sales, Productions, and Purchases (Europe) (*continued*)

Competitive Abilities	Industry	
	Total	Consumer
Domestic purchases, 1992		
Mean	82.63	82.11
Standard deviation	21.55	26.48
Number	171	35
Foreign purchases, 1992		
Mean	17.11	17.94
Standard deviation	21.12	26.59
Number	171	35

Industry			
Industrial Goods	*Basic*	*Machinery*	*Electronics*
91.19	82.32	89.39	73.78*
10.96	22.98	10.92	22.37
16	50	33	37
8.81	17.48	10.73	25.11*
10.96	23.05	10.92	20.61
16	50	33	37

*The ANOVA F-test calculated from the comparison between industries within Europe is significant at the .05 alpha level.

(concluded)

Business Unit Growth Strategy by Industry (Japan)

	Industry		
Indicators	Total	Consumer	Industrial Goods
Build market share			
Mean	0.63	0.67	N/A
Standard deviation	0.48	0.50	N/A
Number	133	9	N/A
Hold (defend) market share			
Mean	0.33	0.33	N/A
Standard deviation	0.47	0.50	N/A
Number	133	9	N/A
Harvest (maximize cash flow, sacrifice market share)			
Mean	0.03	0.00	N/A
Standard deviation	0.17	0.00	N/A
Number	133	9	N/A
Withdraw (prepare to exit the business)			
Mean	0.00	0.00	N/A
Standard deviation	0.00	0.00	N/A
Number	133	9	N/A

	Industry	
Basic	Machinery	Electronics
0.56	0.62	0.70
0.50	0.49	0.46
34	47	43
0.44	0.34	0.23
0.50	0.48	0.43
34	47	43
0.00	0.02	0.07
0.00	0.15	0.26
34	47	43
0.00	0.00	0.00
0.00	0.00	0.00
34	47	43

Selected Business Unit Profile by Industry (Japan)

	Industry	
Indicators	Total	Consumer
Annual sales revenue[†]		
Mean	1,258.95	783.29
Standard deviation	3,910.43	813.85
Number	115	7
Pre-tax return on assets (% of assets)		
Mean	10.70	19.80
Standard deviation	13.10	14.25
Number	74	8
Net pre-tax profit ratio (% of sales)		
Mean	14.60	5.00
Standard deviation	65.07	2.00
Number	87	5
R&D expenses (% of sales)		
Mean	3.60	2.67
Standard deviation	2.84	0.58
Number	83	3
Growth rate in unit sales (%)		
Mean	54.61	88.75
Standard deviation	83.47	58.13
Number	110	8
Market share of primary product		
Mean	27.84	26.29
Standard deviation	17.59	14.84
Number	103	7
Capacity utilization (%)		
Mean	98.30	95.50
Standard deviation	19.30	14.43
Number	107	8
Number of plants in the business unit		
Mean	3.37	5.50
Standard deviation	4.46	6.65
Number	111	8
Number of employees		
Mean	2,090.92	1,065.89
Standard deviation	5,812.72	1,586.23
Number	128	9
Manufacturing direct labor employees		
Mean	1,161.96	580.56
Standard deviation	2,741.12	715.77
Number	116	9

	Industry		
Industrial Goods	*Basic*	*Machinery*	*Electronics*
N/A	1,209.14	1,838.15	797.85
	1,771.33	6,375.66	1,081.91
	28	40	40
N/A	10.80	6.44	13.22
	15.15	4.09	16.39
	15	27	27
N/A	9.50	24.88	8.16
	18.54	104.81	6.01
	18	33	31
N/A	4.06	2.63	4.34
	3.15	1.77	3.36
	18	30	32
N/A	68.04	42.82	50.03
	146.83	45.60	51.92
	26	38	38
N/A	18.64	32.31	30.14*
	14.05	20.42	15.22
	25	36	35
N/A	94.89	101.42	98.18
	26.28	20.10	11.51
	27	38	34
N/A	5.00	2.39	2.77*
	7.45	1.53	2.31
	27	41	35
N/A	1,233.36	3,304.22	1,674.49
	1,824.52	9,356.53	2,265.74
	33	45	41
N/A	874.66	1,783.84	785.49
	1,696.51	4,122.93	1,063.09
	29	43	35

Selected Business Unit Profile by Industry (Japan) (*continued*)

Indicators	Industry	
	Total	*Consumer*
Manufacturing indirect labor employees		
Mean	1,007.66	229.56
Standard deviation	3,498.18	249.49
Number	116	9

Industry			
Industrial Goods	*Basic*	*Machinery*	*Electronics*
N/A	418.90	1,644.19	913.57
	454.21	5,532.35	155.74
	29	43	35

†In U.S. dollars.
*The ANOVA F-test calculated from the comparison between industries within Japan is significant at the .05 alpha level.

(*concluded*)

Importance of Competitive Abilities by Industry (Japan)

	Industry	
Importance of Competitive Abilities	Total	Consumer
Ability to profit in price-competitive market		
Mean	5.48	5.33
Standard deviation	1.26	1.32
Number	132	9
Ability to make rapid changes in design		
Mean	5.84	5.67
Standard deviation	1.10	1.00
Number	132	9
Ability to introduce new products quickly		
Mean	5.84	5.67
Standard deviation	1.10	1.00
Number	132	9
Ability to make rapid volume changes		
Mean	5.35	5.22
Standard deviation	1.19	1.09
Number	132	9
Ability to make rapid product mix changes		
Mean	5.03	5.22
Standard deviation	1.34	0.97
Number	125	9
Ability to offer a broad product line		
Mean	5.16	5.44
Standard deviation	1.35	1.01
Number	132	9
Ability to offer consistently low defect rates		
Mean	5.85	5.44
Standard deviation	1.15	1.74
Number	131	9
Ability to provide high-performance products or product amenities		
Mean	5.42	5.33
Standard deviation	1.25	1.50
Number	131	9
Ability to provide reliable/durable products		
Mean	6.20	6.22
Standard deviation	0.92	0.83
Number	131	9
Ability to provide fast deliveries		
Mean	5.61	5.67
Standard deviation	1.18	1.22
Number	132	9

	Industry		
Industrial Goods	Basic	Machinery	Electronics
N/A	5.44	5.50	5.53
	1.24	1.38	1.16
	34	46	43
N/A	5.85	6.00	5.70
	1.18	0.94	1.23
	34	46	43
N/A	5.85	6.00	5.70
	1.18	.94	1.23
	34	46	43
N/A	5.56	5.28	5.28
	0.96	1.38	1.16
	34	46	43
N/A	4.71	5.18	5.08
	1.66	1.34	1.12
	31	45	40
N/A	5.44	5.20	4.84
	1.33	1.38	1.36
	34	46	43
N/A	6.09	5.84	5.74
	0.93	1.26	1.05
	34	45	43
N/A	5.27	5.57	5.40
	1.44	1.11	1.20
	33	46	43
N/A	6.15	6.17	6.26
	0.96	1.06	0.77
	34	46	42
N/A	5.59	5.61	5.63
	1.23	1.24	1.11
	34	46	43

Importance of Competitive Abilities by Industry (Japan) (*continued*)

	Industry	
Importance of Competitive Abilities	*Total*	*Consumer*
Ability to make dependable delivery promises		
Mean	5.97	6.00
Standard deviation	1.05	1.00
Number	131	9
Ability to provide effective after-sales service		
Mean	5.52	5.67
Standard deviation	1.12	0.87
Number	132	9
Ability to provide product support effectively		
Mean	5.43	5.11
Standard deviation	1.08	0.93
Number	132	9
Ability to make products easily available (broad distribution)		
Mean	5.04	5.11
Standard deviation	1.34	1.17
Number	129	9
Ability to customize products and services to customer needs		
Mean	5.88	5.89
Standard deviation	1.00	1.05
Number	130	9

	Industry		
Industrial Goods	*Basic*	*Machinery*	*Electronics*
N/A	6.06	6.07	5.79
	1.00	1.00	1.17
	33	46	43
N/A	5.29	5.80	5.37
	1.36	0.91	1.13
	34	46	43
N/A	5.47	5.63	5.26
	1.19	1.14	0.93
	34	46	43
N/A	5.21	5.09	4.83
	1.30	1.47	1.26
	34	45	41
N/A	6.03	5.98	5.67
	0.98	0.83	1.16
	33	46	42

(concluded)

Strength in Competitive Abilities by Industry (Japan)

	Industry	
Competitive Abilities	Total	Consumer
Ability to profit in price-competitive market		
Mean	4.09	4.13
Standard deviation	1.34	1.73
Number	132	8
Ability to make rapid changes in design		
Mean	4.36	4.13
Standard deviation	1.20	1.64
Number	131	8
Ability to introduce new products quickly		
Mean	4.36	4.13
Standard deviation	1.20	1.64
Number	131	8
Ability to make rapid volume changes		
Mean	4.62	4.00
Standard deviation	1.11	0.53
Number	132	8
Ability to make rapid product mix changes		
Mean	4.62	4.00
Standard deviation	1.11	0.53
Number	132	8
Ability to offer a broad product line		
Mean	4.46	5.13
Standard deviation	1.29	1.36
Number	130	8
Ability to offer consistently low defect rates		
Mean	4.91	5.25
Standard deviation	1.13	1.28
Number	131	8
Ability to provide high-performance products or product amenities		
Mean	4.43	4.63
Standard deviation	1.17	1.19
Number	131	8
Ability to provide reliable/durable products		
Mean	5.06	4.88
Standard deviation	1.09	1.13
Number	130	8
Ability to provide fast deliveries		
Mean	4.48	4.25
Standard deviation	1.16	1.28
Number	132	8

	Industry		
Industrial Goods	Basic	Machinery	Electronics
N/A	4.24	4.02	4.05
	1.39	1.42	1.15
	34	47	43
N/A	4.29	4.43	4.38
	1.43	1.04	1.13
	34	47	42
N/A	4.29	4.43	4.38
	1.43	1.04	1.13
	34	47	42
N/A	4.62	4.64	4.72
	1.02	1.24	1.10
	34	47	43
N/A	4.62	4.64	4.72
	1.02	1.24	1.10
	34	47	43
N/A	4.67	4.28	4.38
	1.22	1.39	1.19
	33	47	42
N/A	5.03	4.80	4.86
	1.14	1.17	1.06
	34	46	43
N/A	4.18	4.51	4.49
	1.13	1.18	1.20
	33	47	43
N/A	5.06	5.04	5.12
	1.15	1.11	1.04
	34	46	42
N/A	4.65	4.45	4.44
	1.01	1.36	1.01
	34	47	43

Strength in Competitive Abilities by Industry (Japan) (*continued*)

Competitive Abilities	Industry	
	Total	Consumer
Ability to make dependable delivery promises		
Mean	4.82	4.75
Standard deviation	1.29	1.17
Number	132	8
Ability to provide effective after-sales service		
Mean	4.66	4.88
Standard deviation	1.18	1.25
Number	132	8
Ability to provide product support effectively		
Mean	4.59	4.50
Standard deviation	1.13	1.07
Number	132	8
Ability to make products easily available (broad distribution)		
Mean	4.53	5.25
Standard deviation	1.10	1.04
Number	130	8
Ability to customize products and services to customer needs		
Mean	4.71	4.75
Standard deviation	1.15	1.17
Number	129	8

	Industry		
Industrial Goods	Basic	Machinery	Electronics
N/A	5.06	4.66	4.81
	1.18	1.48	1.18
	34	47	43
N/A	4.59	4.47	4.88
	1.16	1.16	1.22
	34	47	43
N/A	4.79	4.49	4.56
	0.98	1.12	1.26
	34	47	43
N/A	4.56	4.48	4.43
	1.19	1.13	0.99
	34	46	42
N/A	4.75	4.62	4.76
	1.24	1.19	1.05
	32	47	42

(concluded)

Manufacturing Cost Structure by Industry (Japan)

	Industry	
Manufacturing Cost Structure	Total	Consumer
Manufacturing costs as percent of sales, 1989		
Mean	76.20	70.57
Standard deviation	13.02	20.29
Number	108	7
Manufacturing costs as percent of sales, 1992		
Mean	73.55	67.00
Standard deviation	12.65	23.61
Number	88	5
Materials cost (% of total mfg. cost), 1989*		
Mean		
Standard deviation		
Number		
Materials cost (% of total mfg. cost), 1992*		
Mean		
Standard deviation		
Number		
Direct labor cost (% of total mfg. cost), 1989*		
Mean		
Standard deviation		
Number		
Direct labor cost (% of total mfg. cost), 1992*		
Mean		
Standard deviation		
Number		
Energy cost (% of total mfg. cost), 1989*		
Mean		
Standard deviation		
Number		
Energy cost (% of total mfg. cost), 1992*		
Mean		
Standard deviation		
Number		
Manufacturing overhead cost (% of total mfg. cost), 1989*		
Mean		
Standard deviation		
Number		
Manufacturing overhead cost (% of total mfg. cost), 1992*		
Mean		
Standard deviation		
Number		

| | Industry | | |
Industrial Goods	Basic	Machinery	Electronics
N/A	75.12	81.44	71.97[†]
	14.08	9.69	12.29
	25	41	35
N/A	74.84	77.00	69.90
	12.31	10.26	12.37
	19	34	30

*The Japanese survey instrument did not solicit responses to these items.

[†]The ANOVA F-test calculated from the comparison between industries within Japan is significant at the .05 alpha level.

Manufacturing Performance Improvement Index by Industry (Japan)

Performance Indicators	Industry	
	Total	Consumer
Overall quality as perceived by customers		
Mean	110.72	106.11
Standard deviation	14.71	12.19
Number	120	9
Average unit production costs for typical product		
Mean	105.24	102.56
Standard deviation	12.40	9.62
Number	123	9
Inventory turnover		
Mean	113.83	113.67
Standard deviation	30.47	29.26
Number	124	9
Speed of new product development and/or design change		
Mean	108.82	107.78
Standard deviation	15.35	9.05
Number	121	9
On-time delivery		
Mean	108.79	103.33
Standard deviation	25.44	6.61
Number	123	9
Equipment changeover (or set-up) time		
Mean	112.70	108.89
Standard deviation	16.57	13.41
Number	120	9
Market share		
Mean	103.59	103.50
Standard deviation	10.37	4.72
Number	119	8
Profitability		
Mean	118.54	111.25
Standard deviation	51.47	26.42
Number	114	8
Customer service		
Mean	108.06	107.78
Standard deviation	11.54	19.70
Number	116	9
Manufacturing lead time		
Mean	108.01	105.56
Standard deviation	13.26	12.86
Number	122	9

	Industry		
Industrial Goods	Basic	Machinery	Electronics
N/A	101.56	112.91	109.45
	11.79	13.60	17.93
	27	44	40
N/A	103.31	104.51	108.08
	14.27	9.59	14.09
	29	45	40
N/A	107.38	116.44	115.56
	17.84	35.64	31.97
	29	45	41
N/A	104.36	108.98	112.00
	8.24	12.97	21.24
	28	44	40
N/A	109.24	109.82	108.53
	11.43	31.86	27.72
	29	45	40
N/A	113.11	114.68	111.05
	13.94	19.87	15.04
	28	44	39
N/A	106.93	101.23	103.72
	12.54	6.90	12.12
	29	43	39
N/A	107.36	114.05	132.05
	19.17	34.42	77.04
	25	42	39
N/A	107.96	108.34	107.86
	9.53	9.48	13.05
	27	44	36
N/A	105.76	108.42	109.77
	17.44	10.27	13.05
	29	45	39

Manufacturing Performance Improvement Index by Industry (Japan) (*continued*)

	Industry	
Performance Indicators	*Total*	*Consumer*
Procurement lead time		
Mean	105.31	104.44
Standard deviation	11.47	11.02
Number	121	9
Delivery lead time		
Mean	108.46	105.56
Standard deviation	12.87	14.46
Number	120	9
Variety of products producible by manufacturing		
Mean	115.37	126.11
Standard deviation	21.91	43.72
Number	121	9

	Industry		
Industrial Goods	Basic	Machinery	Electronics
N/A	105.41	104.95	105.85
	11.46	12.58	10.64
	29	44	39
N/A	107.93	107.78	110.41
	12.07	12.66	13.61
	29	45	37
N/A	112.46	111.02	120.00
	12.01	13.21	27.05
	28	45	39

(*concluded*)

Manufacturing Objectives by Industry (Japan)

Manufacturing's Objectives	Industry	
	Total	Consumer
Improve conformance quality (reduce defects)		
Mean	6.05	5.78
Standard deviation	1.05	1.39
Number	133	9
Reduce unit costs		
Mean	6.23	6.22
Standard deviation	0.92	0.97
Number	133	9
Improve safety record		
Mean	5.39	5.22
Standard deviation	1.34	0.83
Number	131	9
Reduce manufacturing lead time		
Mean	5.79	5.78
Standard deviation	0.98	0.97
Number	133	9
Increase capacity		
Mean	5.47	5.22
Standard deviation	1.16	0.97
Number	133	9
Reduce procurement lead time		
Mean	5.45	5.11
Standard deviation	1.13	1.27
Number	132	9
Reduce new product development cycle		
Mean	5.70	5.44
Standard deviation	1.33	1.24
Number	131	9
Reduce materials costs		
Mean	5.51	4.67
Standard deviation	1.13	1.00
Number	131	9
Reduce overhead costs		
Mean	5.37	4.67
Standard deviation	1.14	0.87
Number	131	9
Improve direct labor productivity		
Mean	5.91	5.78
Standard deviation	0.92	0.97
Number	132	9

	Industry		
Industrial Goods	*Basic*	*Machinery*	*Electronics*
N/A	6.06	6.19	5.95
	1.20	0.97	0.92
	34	47	43
N/A	6.12	6.23	6.30
	1.15	0.87	0.77
	34	47	43
N/A	5.88	5.76	4.65*
	1.27	1.14	1.38
	33	46	43
N/A	5.53	5.79	6.00
	1.08	0.93	0.93
	34	47	43
N/A	5.74	5.34	5.47
	1.29	1.18	1.05
	34	47	43
N/A	4.79	5.60	5.86*
	1.29	0.97	0.92
	33	47	43
N/A	5.18	5.87	5.98*
	1.38	1.44	1.09
	33	47	42
N/A	5.48	5.55	5.67
	1.25	1.16	0.98
	33	47	42
N/A	5.64	5.26	5.43
	1.27	1.19	0.99
	33	47	42
N/A	6.09	5.81	5.90
	1.08	0.82	0.88
	34	47	42

Manufacturing Objectives by Industry (Japan) (*continued*)

	Industry	
Manufacturing's Objectives	*Total*	*Consumer*
Increase throughput		
Mean	5.80	5.67
Standard deviation	0.96	1.00
Number	131	9
Reduce number of vendors		
Mean	3.81	3.78
Standard deviation	1.37	1.09
Number	131	9
Improve vendor quality		
Mean	5.31	5.00
Standard deviation	1.09	1.00
Number	131	9
Reduce inventories		
Mean	5.35	5.78
Standard deviation	1.17	0.97
Number	133	9
Increase delivery reliability		
Mean	5.58	5.44
Standard deviation	1.18	0.88
Number	132	9
Increase delivery speed		
Mean	5.56	5.33
Standard deviation	1.03	0.71
Number	133	9
Improve ability to make rapid product mix changes		
Mean	5.42	4.89
Standard deviation	0.96	0.78
Number	130	9
Improve ability to make rapid volume changes		
Mean	5.13	5.11
Standard deviation	1.09	0.93
Number	132	9
Reduce break-even points		
Mean	5.88	5.78
Standard deviation	1.04	1.09
Number	132	9
Raise employee morale		
Mean	5.46	5.44
Standard deviation	1.06	0.73
Number	133	9

| | Industry | | |
Industrial Goods	Basic	Machinery	Electronics
N/A	5.94	5.74	5.79
	1.00	0.97	0.95
	33	47	42
N/A	3.73	3.64	4.07
	1.46	1.42	1.30
	33	47	42
N/A	4.94	5.57	5.36
	1.17	1.08	0.98
	33	47	42
N/A	5.35	5.04	5.58
	1.35	1.25	0.91
	34	47	43
N/A	5.48	5.74	5.51
	1.46	1.22	0.94
	33	47	43
N/A	5.47	5.64	5.60
	1.08	1.11	0.98
	34	47	43
N/A	5.18	5.41	5.74*
	0.92	0.91	0.99
	33	46	42
N/A	5.03	5.19	5.14
	0.98	1.08	1.25
	33	47	43
N/A	5.94	5.87	5.86
	1.03	0.99	1.13
	33	47	43
N/A	5.38	5.70	5.26
	1.13	1.02	1.09
	34	47	43

Manufacturing Objectives by Industry (Japan) (*continued*)

Manufacturing's Objectives	Industry	
	Total	Consumer
Maximize cash flow		
Mean	5.12	4.88
Standard deviation	0.96	0.64
Number	125	8
Increase environmental safety and protection		
Mean	5.20	5.56
Standard deviation	1.29	1.13
Number	131	9
Reduce capacity		
Mean	2.06	1.78
Standard deviation	1.21	0.83
Number	131	9
Increase product or materials standardization		
Mean	4.98	4.89
Standard deviation	1.30	1.05
Number	131	9
Improve labor relations		
Mean	4.98	5.44
Standard deviation	1.08	0.73
Number	131	9
Improve white-collar productivity		
Mean	5.27	4.89
Standard deviation	1.14	0.93
Number	131	9
Increase range of products produced by existing facilities		
Mean	4.43	4.44
Standard deviation	1.23	1.33
Number	131	9
Meet financial shipping goals		
Mean	4.43	3.89
Standard deviation	1.31	0.93
Number	130	9
Improve pre-sales service and technical support		
Mean	4.95	4.56
Standard deviation	1.06	0.88
Number	128	9
Improve after-sales service		
Mean	5.00	4.78
Standard deviation	1.03	0.67
Number	129	9

	Industry		
Industrial Goods	**Basic**	**Machinery**	**Electronics**
N/A	5.00	5.20	5.18
	1.15	0.96	0.84
	31	46	40
N/A	5.61	5.19	4.81*
	1.20	1.35	1.25
	33	47	42
N/A	1.85	2.30	2.02
	1.23	1.38	1.02
	33	47	42
N/A	4.48	5.02	5.33*
	1.33	1.31	1.22
	33	47	42
N/A	4.61	5.17	4.98
	1.06	1.13	1.05
	33	47	42
N/A	5.03	5.51	5.29
	1.29	0.88	1.29
	33	47	42
N/A	4.36	4.60	4.29
	1.43	1.15	1.13
	33	47	42
N/A	4.36	4.46	4.57
	1.11	1.38	1.43
	33	46	42
N/A	4.82	5.20	4.88
	1.13	1.00	1.16
	33	46	40
N/A	4.97	5.21	4.83
	1.02	0.98	1.15
	33	47	40

Manufacturing Objectives by Industry (Japan) (*continued*)

	Industry	
Manufacturing's Objectives	*Total*	*Consumer*
Change culture of manufacturing organization		
Mean	5.02	4.78
Standard deviation	1.08	1.09
Number	130	9
Improve interfunctional communication		
Mean	5.19	5.33
Standard deviation	1.04	1.22
Number	131	9
Improve communication with external partners		
Mean	5.10	5.00
Standard deviation	1.01	1.12
Number	131	9
Reduce set-up/changeover times		
Mean	5.53	5.67
Standard deviation	1.04	0.50
Number	131	9

	Industry		
Industrial Goods	Basic	Machinery	Electronics
N/A	4.85	5.26	4.95
	1.12	0.93	1.19
	33	46	42
N/A	5.27	5.30	4.98
	1.10	0.88	1.12
	33	47	42
N/A	5.00	5.11	5.19
	1.15	0.76	1.13
	33	47	42
N/A	5.52	5.64	5.38
	1.25	0.97	1.03
	33	47	42

*The ANOVA F-test calculated from the comparison between industries within Japan is significant at the .05 alpha level.

(*concluded*)

Past Pay-offs from Programs and Activities by Industry (Japan)

	Industry	
Programs / Activities	Total	Consumer
Giving workers a broad range of tasks and/or more responsibility		
Mean	4.39	4.00
Standard deviation	1.24	1.32
Number	122	9
Activity-based costing		
Mean	3.61	3.75
Standard deviation	1.48	1.58
Number	111	8
Manufacturing reorganization		
Mean	4.27	3.44
Standard deviation	1.37	1.81
Number	115	9
Worker training		
Mean	4.43	4.00
Standard deviation	1.17	1.41
Number	116	9
Management training		
Mean	4.62	4.88
Standard deviation	1.18	1.25
Number	116	8
Supervisor training		
Mean	4.46	4.63
Standard deviation	1.16	1.30
Number	115	8
Computer-aided manufacturing (CAM)		
Mean	4.10	3.00
Standard deviation	1.67	1.83
Number	110	7
Computer-aided design (CAD)		
Mean	4.91	3.13
Standard deviation	1.55	1.89
Number	120	8
Value analysis/product redesign		
Mean	4.74	3.17
Standard deviation	1.39	1.72
Number	117	6
Interfunctional work teams		
Mean	4.37	4.33
Standard deviation	1.08	0.82
Number	111	6

	Industry		
Industrial Goods	Basic	Machinery	Electronics
N/A	4.10	4.68	4.38
	1.09	1.27	1.27
	30	44	39
N/A	3.34	3.65	3.76
	1.42	1.60	1.40
	29	37	37
N/A	4.34	4.41	4.26
	1.17	1.45	1.29
	29	39	38
N/A	4.48	4.62	4.28
	1.21	1.23	0.97
	29	42	36
N/A	4.73	4.51	4.59
	1.14	1.27	1.14
	30	41	37
N/A	4.67	4.54	4.17
	1.18	1.29	0.94
	30	41	36
N/A	3.89	4.15	4.43
	1.45	1.67	1.75
	28	40	35
N/A	4.00	5.32	5.49*
	1.63	1.33	1.05
	29	44	39
N/A	4.35	5.10	4.92*
	1.23	1.18	1.48
	31	41	39
N/A	4.45	4.39	4.29
	1.21	1.18	0.89
	29	41	35

Past Pay-offs from Programs and Activities by Industry (Japan) (*continued*)

	Industry	
Programs / Activities	*Total*	*Consumer*
Quality function deployment		
Mean	4.83	4.57
Standard deviation	1.25	1.51
Number	111	7
Developing new processes for new products		
Mean	4.89	5.56
Standard deviation	1.23	0.88
Number	118	9
Developing new processes for old products		
Mean	5.08	5.33
Standard deviation	1.17	1.00
Number	119	9
Integrating information systems in manufacturing		
Mean	4.54	4.13
Standard deviation	1.29	1.46
Number	119	8
Integrating information systems across functions		
Mean	4.30	3.50
Standard deviation	1.34	1.07
Number	115	8
Reconditioning physical plants		
Mean	4.10	4.00
Standard deviation	1.29	1.85
Number	106	8
Just-in-time		
Mean	4.26	3.75
Standard deviation	1.63	1.98
Number	117	8
Robots		
Mean	4.27	3.11
Standard deviation	1.63	1.62
Number	116	9
Flexible manufacturing systems		
Mean	4.01	4.13
Standard deviation	1.31	0.64
Number	114	8
Design for manufacture		
Mean	4.30	4.00
Standard deviation	1.15	0.58
Number	110	7

Industrial Goods	Basic	Machinery	Electronics
	Industry		
N/A	4.93	4.80	4.82
	1.01	1.38	1.27
	30	40	34
N/A	5.27	4.48	4.87*
	1.28	1.18	1.20
	30	40	39
N/A	5.30	4.98	4.95
	1.15	1.27	1.13
	30	40	40
N/A	4.76	4.53	4.46
	1.18	1.33	1.31
	29	43	39
N/A	4.32	4.36	4.38
	4.52	1.16	1.41
	28	39	40
N/A	4.15	4.17	4.03
	1.05	1.27	1.38
	26	35	37
N/A	4.13	4.46	4.26
	1.63	1.65	1.57
	31	39	39
N/A	1.62	4.64	4.20
	1.74	1.39	1.71
	30	42	35
N/A	3.72	4.18	4.03
	1.22	1.32	1.48
	29	40	37
N/A	4.04	4.39	4.46
	1.32	1.09	1.15
	27	41	35

**Past Pay-offs from Programs and Activities by Industry
(Japan)** (*continued*)

	Industry	
Programs / Activities	*Total*	*Consumer*
Statistical quality control		
Mean	4.42	4.38
Standard deviation	1.12	1.06
Number	110	8
Closing and/or relocating plants		
Mean	2.58	3.88
Standard deviation	1.98	2.03
Number	105	8
Quality circles		
Mean	4.86	4.88
Standard deviation	1.25	1.55
Number	119	8
Investing in improved production-inventory control systems		
Mean	4.60	4.50
Standard deviation	1.32	0.93
Number	121	8
Hiring in new skills from outside		
Mean	3.79	3.63
Standard deviation	1.28	1.51
Number	108	8
Linking manufacturing strategy to business strategy		
Mean	4.44	4.75
Standard deviation	1.39	0.89
Number	111	8

| | Industry | | |
Industrial Goods	Basic	Machinery	Electronics
N/A	4.66	4.19	4.47
	1.26	1.05	1.08
	29	37	36
N/A	1.89	2.64	2.79
	1.71	1.99	2.04
	28	36	33
N/A	4.91	4.48	5.21
	1.40	1.24	0.96
	33	40	38
N/A	4.75	4.40	4.71
	1.32	1.38	1.33
	32	43	38
N/A	3.79	3.76	3.85
	1.29	1.26	1.28
	28	38	34
N/A	4.64	4.26	4.41
	1.52	1.29	1.50
	28	38	37

*The ANOVA F-test calculated from the comparison between industries within Japan is significant at the .05 alpha level.

(concluded)

Future Emphasis on Programs and Activities by Industry (Japan)

	Industry	
Programs / Activities	Total	Consumer
Giving workers a broad range of tasks and/or more responsibility		
Mean	5.03	4.44
Standard deviation	1.17	1.24
Number	119	9
Activity-based costing		
Mean	4.28	0.50
Standard deviation	1.42	1.41
Number	107	8
Manufacturing reorganization		
Mean	4.68	4.00
Standard deviation	1.17	1.66
Number	111	9
Worker training		
Mean	5.01	4.89
Standard deviation	1.05	1.17
Number	114	9
Management training		
Mean	5.34	5.00
Standard deviation	1.10	1.31
Number	115	8
Supervisor training		
Mean	5.15	5.00
Standard deviation	1.10	1.07
Number	133	8
Computer-aided manufacturing (CAM)		
Mean	5.08	3.71
Standard deviation	1.69	2.36
Number	110	7
Computer-aided design (CAD)		
Mean	5.41	3.63
Standard deviation	1.36	1.77
Number	119	8
Value analysis/product redesign		
Mean	5.29	4.33
Standard deviation	1.28	2.25
Number	113	6
Interfunctional work teams		
Mean	4.71	4.14
Standard deviation	1.24	2.04
Number	107	7

	Industry		
Industrial Goods	**Basic**	**Machinery**	**Electronics**
N/A	4.97	5.07	5.15
	1.15	1.21	1.12
	29	41	40
N/A	4.29	4.49	4.24
	1.27	1.61	1.30
	28	37	34
N/A	4.33	4.87	4.89
	1.04	0.96	1.24
	27	38	37
N/A	5.04	5.08	4.95
	1.04	1.00	1.13
	28	40	37
N/A	5.47	5.21	5.45
	1.04	1.08	1.13
	30	39	38
N/A	5.14	5.15	5.19
	1.08	1.09	1.17
	28	41	36
N/A	4.85	5.41	5.16
	1.63	1.37	1.82
	27	39	37
N/A	4.86	5.74	5.80*
	1.56	1.05	1.02
	28	43	40
N/A	4.79	5.67	5.44*
	1.24	1.13	1.12
	29	39	39
N/A	4.67	4.72	4.85
	1.07	1.23	1.21
	27	39	34

Future Emphasis on Programs and Activities by Industry (Japan) (*continued*)

	Industry	
Programs / Activities	Total	Consumer
Quality function deployment		
Mean	5.35	5.43
Standard deviation	1.19	0.79
Number	109	7
Developing new processes for new products		
Mean	5.75	5.89
Standard deviation	1.05	0.93
Number	116	9
Developing new processes for old products		
Mean	5.53	6.00
Standard deviation	1.11	1.00
Number	117	9
Integrating information systems in manufacturing		
Mean	5.78	5.63
Standard deviation	1.06	1.19
Number	118	8
Integrating information systems across functions		
Mean	5.51	5.13
Standard deviation	1.15	1.25
Number	114	8
Reconditioning physical plants		
Mean	4.86	4.75
Standard deviation	1.38	1.04
Number	107	8
Just-in-time		
Mean	4.90	5.13
Standard deviation	1.55	1.36
Number	114	8
Robots		
Mean	5.10	4.11
Standard deviation	1.44	1.45
Number	114	9
Flexible manufacturing systems		
Mean	5.04	5.00
Standard deviation	1.55	1.20
Number	133	8
Design for manufacture		
Mean	5.30	4.71
Standard deviation	1.11	1.11
Number	112	7

	Industry		
Industrial Goods	*Basic*	*Machinery*	*Electronics*
N/A	5.43	5.26	5.37
	1.14	1.35	1.14
	28	39	35
N/A	5.82	5.51	5.90
	1.16	1.12	0.90
	28	39	40
N/A	6.03	5.40	5.18*
	0.91	1.10	1.14
	29	40	39
N/A	5.86	5.76	5.78
	1.08	1.10	1.00
	28	42	40
N/A	5.36	5.56	5.64
	1.37	1.10	1.01
	28	39	39
N/A	4.61	5.12	4.84
	1.50	1.51	1.24
	28	34	37
N/A	4.83	4.70	5.10
	1.64	1.65	1.45
	30	37	39
N/A	5.03	5.51	4.91*
	1.49	1.49	1.20
	29	41	35
N/A	4.68	5.31	5.03
	1.87	1.61	1.28
	28	39	38
N/A	5.21	5.43	5.36
	1.08	1.13	1.13
	29	40	36

Future Emphasis on Programs and Activities by Industry (Japan) (*continued*)

	Industry	
Programs / Activities	Total	Consumer
Statistical quality control		
Mean	4.71	4.50
Standard deviation	1.17	1.07
Number	110	8
Closing and/or relocating plants		
Mean	2.84	4.75
Standard deviation	2.05	1.75
Number	106	8
Quality circles		
Mean	5.00	5.25
Standard deviation	1.22	1.58
Number	117	8
Investing in improved production-inventory control systems		
Mean	5.61	5.63
Standard deviation	1.11	0.52
Number	121	8
Hiring in new skills from outside		
Mean	4.71	4.38
Standard deviation	1.29	0.92
Number	109	8
Linking manufacturing strategy to business strategy		
Mean	5.57	5.63
Standard deviation	1.20	1.06
Number	112	8

	Industry		
Industrial Goods	*Basic*	*Machinery*	*Electronics*
N/A	5.04	4.51	4.70
	1.32	1.15	1.08
	28	37	37
N/A	2.21	2.64	3.12*
	1.97	1.91	2.07
	28	36	34
N/A	5.00	4.83	5.14
	1.39	1.18	1.06
	31	41	37
N/A	5.77	5.50	5.60
	1.15	1.15	1.13
	31	42	40
N/A	4.54	4.84	4.77
	1.35	1.26	1.35
	28	38	35
N/A	5.50	5.67	5.51
	1.37	1.15	1.17
	28	39	37

*The ANOVA F-test calculated from the comparison between industries within Japan is significant at the .05 alpha level.

(*concluded*)

Internal Environment by Industry (Japan)

	Industry		
Indicators	Total	Consumer	Industrial Goods
Top management only			
Mean	0.01	0.00	N/A
Standard deviation	0.09	0.00	N/A
Number	133	9	N/A
Top and some middle management			
Mean	0.10	0.11	N/A
Standard deviation	0.30	0.33	N/A
Number	133	9	N/A
Top and most middle management			
Mean	0.35	0.56	N/A
Standard deviation	0.48	0.53	N/A
Number	133	9	N/A
Every manager and supervisor			
Mean	0.35	0.22	N/A
Standard deviation	0.48	0.44	N/A
Number	133	9	N/A
Every manager, supervisor, and worker			
Mean	0.18	0.11	N/A
Standard deviation	0.39	0.33	N/A
Number	133	9	N/A

	Industry	
Basic	Machinery	Electronics
0.03	0.00	0.00
0.17	0.00	0.00
34	47	43
0.03	0.19	0.05
0.17	0.40	0.21
34	47	43
0.38	0.26	0.40
0.49	0.44	0.49
34	47	43
0.29	0.34	0.44
0.46	0.48	0.50
34	47	43
0.26	0.19	0.12
0.45	0.40	0.32
34	47	43

Past and Future Percentage of Domestic and Foreign Sales, Productions, and Purchases (Japan)

	Industry	
Competitive Abilities	Total	Consumer
Domestic sales, 1989		
Mean	83.10	90.38
Standard deviation	18.04	17.74
Number	112	8
Foreign sales, 1989		
Mean	17.38	9.75
Standard deviation	18.93	17.67
Number	112	8
Domestic sales, 1992		
Mean	88.30	93.63
Standard deviation	16.93	17.63
Number	111	8
Foreign sales, 1992		
Mean	11.53	6.38
Standard deviation	16.12	17.63
Number	110	8
Domestic production, 1989		
Mean	93.13	99.88
Standard deviation	12.89	0.35
Number	109	8
Foreign production, 1989		
Mean	7.81	0.13
Standard deviation	15.65	0.35
Number	109	8
Domestic production, 1992		
Mean	91.76	96.13
Standard deviation	14.57	6.83
Number	105	8
Foreign production, 1992		
Mean	9.10	3.88
Standard deviation	17.10	6.83
Number	105	8
Domestic purchases, 1989		
Mean	89.21	89.63
Standard deviation	18.40	16.04
Number	105	8
Foreign purchases, 1989		
Mean	11.66	10.38
Standard deviation	20.36	16.04
Number	105	8

	Industry		
Industrial Goods	*Basic*	*Machinery*	*Electronics*
N/A	88.10	82.13	78.50
	16.08	18.52	18.29
	29	39	36
N/A	11.93	19.05	21.64
	16.09	20.92	18.21
	29	39	36
N/A	91.65	85.18	88.16
	16.11	20.13	13.08
	26	40	37
N/A	8.35	14.69	11.54
	16.11	18.19	13.16
	26	39	37
N/A	98.00	92.24	88.75*
	4.30	12.45	16.97
	28	37	36
N/A	2.00	10.46	11.31*
	4.30	19.57	16.95
	28	37	36
N/A	90.46	92.03	91.37
	20.70	11.23	14.39
	24	38	35
N/A	9.54	10.32	8.66
	20.70	18.67	14.38
	24	38	35
N/A	83.54	91.78	90.69
	26.14	15.12	14.54
	26	36	35
N/A	16.46	10.75	9.31
	26.14	21.51	14.54
	26	36	35

Past and Future Percentage of Domestic and Foreign Sales, Productions, and Purchases (Japan) (*continued*)

Competitive Abilities	Industry	
	Total	Consumer
Domestic purchases, 1992		
Mean	59.33	59.00
Standard deviation	38.23	38.30
Number	92	6
Foreign purchases, 1992		
Mean	52.81	54.56
Standard deviation	40.58	38.51
Number	118	9

	Industry		
Industrial Goods	Basic	Machinery	Electronics
N/A	51.88	65.18	57.17
	43.10	39.68	35.32
	16	34	36
N/A	62.04	49.37	50.13
	41.64	42.13	39.18
	26	43	40

*The ANOVA F-test calculated from the comparison between industries within Japan is significant at the .05 alpha level.

(*concluded*)

APPENDIX D

GLOSSARY

activity-based costing (ABC) The use of cost accounting tools that attempt to allocate indirect or overhead costs on the basis of related activities, rather than using surrogate allocation bases such as direct labor or machine hours, floor space, or material costs.

after-sales service Service provided after the sale, such as repair or warranty service, customer problem solving, or follow-up, to ensure the customer is satisfied.

computer-aided design (CAD) The use of computers in interactive engineering drawing and the storage and retrieval of designs.

computer-aided manufacturing (CAM) The use of computers to program, direct, and control production equipment.

design for manufacture (DFM) A general approach to designing products that can be more effectively manufactured. Often used in conjunction with databases. Includes such concepts as design for assembly, design for serviceability, or design for test.

direct labor Labor expended in directly adding value to the product. Often called "touch" labor since direct-labor employees usually physically touch the product or its parts.

high-performance products Products with clearly superior attributes such as taste, styling, speed, comfort, and capacity.

indirect labor Any labor, including supervision and management, that is not direct labor. Overhead activities such as material handling, stockroom, inspection, manufacturing engineering, maintenance, supervision, cost accounting, and personnel are usually included.

interfunctional communication Communications and understandings between manufacturing and other functional areas in a business. For example, between manufacturing and marketing, or manufacturing and finance.

just-in-time (JIT) Applies both to a philosophy of eliminating waste, and to a toolset for pacing and controlling production and vendor deliveries on time, with short notice, and with little or no inventory.

manufacturing lead time (MLT) The cumulative time from the beginning of the production cycle until an item is finally finished. Time spent in inventory as work in process, set-up times, move times, inspection, and order preparation time are included.

manufacturing overhead costs The manufacturing overhead costs that are allocated to unit product costs. It includes the cost of indirect labor as well as indirect purchased services and supplies. It excludes unallocated period costs such as sales and marketing and R&D.

procurement lead time The cumulative time from the beginning of the procurement order cycle (order commitment) until the procured item is delivered. Includes vendor lead time, transportation, and receiving and inspection time.

product amenities Extra product features that enhance the basic product and make it easier to use or more enjoyable. For example, a car radio.

product support Activities that support the customer in the use of a product, such as customer education, information about related products or services, and information hotlines.

quality circles The use of teams of employees to diagnose and solve quality problems. Also includes the use of the work team concept for solving other problems related to productivity improvement, safety, and so on.

quality function deployment (QFD) A set of techniques for determining and communicating customer needs and translating them into product and service design specifications and manufacturing methods.

statistical quality control (SQC) The use of statistical techniques for process control or product inspection. Also includes the use of experimental design techniques for process improvement.

value analysis A systematic approach to simplification and standardization of products so that they provide needed value at minimum cost. Usually applied to existing products to reduce part counts and simplify designs.

INDEX

FORECASTING SYSTEMS FOR OPERATIONS MANAGEMENT
Stephen A. DeLurgio and Carl D. Bhame

Understand and implement practical, theoretically sound, and comprehensive forecasting systems. *Forecasting Systems for Operations Management* will assist you in moving products, materials, and timely information through your organization. It's the most comprehensive treatment of forecasting methods available for automated forecasting systems.
1-55623-040-0 $44.95

COMMON SENSE MANUFACTURING
Becoming a Top Value Competitor
James A. Gardner

Gardner details how you can integrate your manufacturing process and transform your company into a world-class competitor...even if you have limited resources. This common-sense approach to quality and service enhancement pays off in faster employee acceptance and systems applications. Using Gardner's straightforward planning suggestions, you can reduce floor space through better plant layout and gradually eliminate work-in-progress inventory.
1-55623-527-5 $34.95

PURCHASING STRATEGIES FOR TOTAL QUALITY
A Guide to Achieving Continuous Improvement
Greg Hutchins

Shows how companies can use purchasing to help meet and exceed customer demands. Hutchins details how elevating the purchasing person from order taker to a partner in the manufacturing process can be vital to the success of quality improvement efforts. He shows you how to establish continuous improvement strategies with suppliers and infuse quality throughout your organization.
1-55623-380-9 $42.50

COMPUTER INTEGRATED MANUFACTURING
Guidelines and Applications from Industrial Leaders
Steven A. Melnyk and Ram Narasimhan

Melnyk and Narasimhan offer a strategic, management-based approach for understanding and implementing CIM in your operation. A CIM book written from the manager's point of view rather than that of a technical expert, *Computer Integrated Manufacturing* is a clear, easy-to-understand guide. It shows you how to develop a strong link between strategy and CIM structure and build a competitive advantage over your business rivals.
1-55623-538-0 $45.00

MANUFACTURING PLANNING AND CONTROL SYSTEMS
Third Edition
Thomas E. Vollmann, William L. Berry, and D. Clay Whybark

In the Third Edition of *Manufacturing Planning and Control Systems*, state-of-the-art concepts and proven techniques are combined to offer a practical solution to enhancing the manufacturing process. Each of the book's central themes — Master Planning, Material Requirements Planning, Inventory Management, Capacity Management, Production Activity Control, and Just-in-Time — has been updated to reflect the newest ideas and practices.
1-55623-608-5 $55.00